Apollo 8

The NASA Mission Reports

Compiled from the NASA archives & Edited
by Robert Godwin

D1502383

All rights reserved under article two of the Berne Copyright Convention (1971).
We acknowledge the financial support of the Government of Canada through the
Book Publishing Industry Development Program for our publishing activities.
Published by Apogee Books an imprint of Collector's Guide Publishing Inc., Box 62034, Burlington, Ontario, Canada, L7R 4K2
Printed and bound in Canada by Webcom Ltd of Toronto
Apollo 8 - The NASA Mission Reports
by Robert Godwin
ISBN 1-896522-50-5
©1999 Apogee Books
All photos courtesy of NASA
2nd Printing

Introduction

In December of 1968 the world looked as if it might be coming apart at the seams, especially if you were an American. It had been a year of war, rioting and assassinations and for some it felt as though it just couldn't get any worse — and then along came Apollo 8 to save the day.

While America's youth were embroiled in the jungles of far east Asia, President Lyndon Johnson was doing his utmost to nurture the American Space program. He had inherited Kennedy's bold goal of landing a man on the moon before the end of the decade and time was running out.

The National Aeronautics and Space Administration had gradually blossomed into one of the biggest public projects of all time and had seen a string of triumphs drawing a straight line from Alan Shepard's successful sub-orbital flight in a Mercury space capsule right through to the dazzling conclusion of the two man Gemini program, with its long list of space "firsts".

It was now time to attempt the launch of a new three-man spacecraft — Apollo.

In comparison to Mercury and Gemini the Apollo was the Cadillac of space vehicles. It had everything from hot water and hot meals to free floating space in which an astronaut could actually experience some of the fun of zero gravity. There was, however, a lot wrong with the first test vehicle.

In their haste to accomplish the seemingly impossible goal of putting men on the moon, the design had been rushed and a catastrophic combination of errors led to the death of the Apollo 1 crew during a simulation on the launch pad.

A year or so went by and NASA's staff and astronauts decided to memorialise the loss of their crew mates by doing what they would have wanted. Succeed.

The Apollo capsule was redesigned and after five unmanned launches (two of which were on the enormous Saturn V booster) the spacecraft was ready to take men into space.

Even before Apollo 7 (with its crew of veteran Wally Schirra, Don Eisele and Walter Cunningham) had left the ground, an even bolder step was being suggested. Rumour had it that the Soviet Union was about to put men into orbit around the moon and after investing so much effort the men and women of NASA were not about to be upstaged. It was therefore decided to make Apollo 8 the first manned mission to head into deep space. The sheer audacity of this suggestion was probably lost on the general public of the time. Up to that point men had barely been a hundred miles above the Earth and Apollo 7 (although it was in itself a bold undertaking - being the first use of a new spacecraft) was not going to accomplish any altitude records.

Now NASA was proposing to strap three men onto the most complex and powerful piece of machinery ever built and send them hurtling a quarter of a million miles into orbit around the moon. At this point no one had ever been lofted into space by the 363 foot tall Saturn V rocket.

It was going to be a first to be remembered.

On December 21st 1968, veteran astronauts Frank Borman and James Lovell and rookie William Anders were sent to the moon. It was later described by a senior NASA official as the boldest single decision that the agency ever made.

In what seemed like a piece of surrealism from the most fertile minds of science fiction, on Christmas eve 1968, the crew were heard reading an extract from the Book of Genesis from lunar orbit. These three men were the first living creatures from the planet Earth to ever look down onto the far side of the moon with their own eyes, and when they looked back they were the first to see Earthrise above a stark and barren lunar landscape. All three men scrambled for cameras to take a snapshot of the entire human race and its home. To this day the three crew mates still debate who took that unforgettable image.

When that picture was seen in magazines and newspapers around the world for one brief moment there seemed to be pause for thought. All the wars and riots and vendettas were thrust briefly into perspective and the people of the planet Earth were shown to be one species on a small blue gem adrift in a black forbidding sea.

On the successful completion of their mission, splashing down one mile from their target, the crew of Apollo 8 proved that a landing on the moon was indeed a possibility. They also gave the world pause for thought by showing how small and vulnerable the planet Earth really is. One person wrote to the crew, "You saved 1968!"

Although the world more often remembers the names of Armstrong and Aldrin and the flight of Apollo 11, it was Apollo 8, in the most daring and audacious fashion, that many believe paved the way for that later astonishing triumph. It seemed therefore that the details of the flight deserved a place in the commercial marketplace. After exhuming a pile of paper from the NASA archives, I have attempted to collect that information into a format that might be of use to historians, students and general space enthusiasts. In the following pages you will find the text of the original NASA press kit combined with three different mission report documents. These have been supplemented by a small color section sampling some of the startling imagery taken during the mission.

I have made no attempt to explore the fascinating human side of the Apollo 8 mission as this is more than adequately covered in other biographical books such as Jim Lovell's own "Lost Moon" and Andrew Chaikin's "A Man On The Moon". This book is more a tribute to the engineers and team that collaborated to make Kennedy's goal a reality. The sheer wealth of technical expertise that the American aerospace community applied is unprecedented. They proved to everyone that lofty goals are attainable. I hope that the Apollo programme only represents one peak and not the zenith of the American era in space and that you find the contents as inspiring and illuminating as I have.

Robert Godwin
(Editor)

Apollo 8
The NASA Mission Reports

(from the archives of the National Aeronautics and Space Administration)

Apollo 8 Press Kit Index

Apollo 8 Pre-flight Mission Report Index

Apollo 8 Pre-flight Supplemental Report Index

Post Flight Mission Objectives Index

Apollo 8 Press Kit

NATIONAL AERONAUTICS & SPACE ADMINISTRATION

FOR RELEASE: SUNDAY
December 15, 1968
RELEASE NO: 68-208

FIRST MANNED LUNAR ORBIT MISSION

The United States has scheduled its first mission designed to orbit men around the Moon for launch Dec. 21 at 7:51 a.m. EST from the National Aeronautics and Space Administration's John F. Kennedy Space Center, Florida.

The mission, designated Apollo 8, will be the second manned flight in the Apollo program and the first manned flight on the Saturn V rocket, the United States' largest launch vehicle.

Crewmen for Apollo 8 are Spacecraft Commander Frank Borman, Command Module Pilot James A. Lovell, Jr. and Lunar Module Pilot William A. Anders. Backup crew is Commander Neil A. Armstrong, Command Module Pilot Edwin E. Aldrin, Jr. and Lunar Module Pilot Fred W. Haise, Jr.

Apollo 8 is an open-ended mission with the objective of proving the capability of the Apollo command and service modules and the crew to operate at lunar distances. A lunar module will not be carried on Apollo 8 but Lunar Test Article (LTA-B) which is equivalent in weight to a lunar module will be carried as ballast.

The mission will be carried out on a step-by-step "commit point" basis. This means that decisions whether to continue the mission or to return to Earth or to change to an alternate mission will be made before each major maneuver based on the status of the spacecraft systems and crew.

A full duration lunar orbit mission would include 10 orbits around the Moon. Earth landing would take place some 147 hours after launch at 10:51 a.m. EST, Dec. 27.

Earlier developmental Apollo Earth-orbital manned and unmanned flights have qualified all the spacecraft systems including the command module heat shield at lunar return speeds — and the Apollo 7 ten-day failure-free mission in October demonstrated that the spacecraft can operate for the lunar-mission duration.

Apollo 8 will gather data to be used in early development of training, ground simulation and crew inflight procedures for later lunar orbit and lunar landing missions.

The Dec. 21 launch date is at the beginning of the December launch window for lunar flights. These windows hinge upon the Moon's position and lunar surface lighting Conditions at the time the spacecraft arrives at the Moon and upon launch and recovery area conditions. The December window closes Dec. 27. The next comparable window opens Jan. 18 and closes Jan. 24.

The mission will be launched from Complex 39A at the Kennedy Space Center on an azimuth varying from 72 to 108 degrees depending on the launch date and time of day of the launch. The first opportunity calls for liftoff at 7:51 a.m. EST Dec. 21 on an azimuth of 72 degrees. Launch of Apollo 8 will mark the first manned use of the Moonport.

The Saturn V launch vehicle with the Apollo spacecraft on top stands 363 feet tall. The five first-stage engines of Saturn V develop a combined thrust of 7,500,000 pounds at liftoff. At ignition the space vehicle weighs 6,218,558 pounds.

Apollo 8 will be inserted into a 103 nautical mile (119 statute miles, 191 kilometers) Earth orbit.

During the second or third Earth orbit, the Saturn V third-stage engine will restart to place the space vehicle on a path to the Moon. The command and service modules will separate from the third stage and begin the translunar coast period at about 66 hours. A lunar orbit insertion burn with the spacecraft Service propulsion engine will place the spacecraft into a 60 x 170 nm (69 x 196 sm. 111 x 314.8 km) elliptical lunar orbit which later will be circularized at 60 nm (69 sm. 111 km).

The translunar injection burn of the third stage will place the spacecraft on a free-return trajectory, so that if for some reason no further maneuvers are made, Apollo 8 would sweep around the Moon and make a direct entry into the Earth's atmosphere at about 136 hours after liftoff and land in the Atlantic off the west coast of Africa. During the free-return trajectory, corrections may be made using the spacecraft Reaction Control System.

Ten orbits will be made around the Moon while the crew conducts navigation and photography investigations, A transearth injection burn with the Service propulsion engine will bring the spacecraft back to Earth with a direct atmospheric entry in the mid-Pacific about 147 hours after a Dec. 21 launch. Missions beginning later in the window would be of longer duration.

Several alternate mission plans are available if for some reason the basic lunar orbit cannot be flown. The alternates range from ten days in low Earth orbit, a high ellipse orbit, to a circumlunar flight with direct Earth entry.

As Apollo 8 leaves Earth orbit and starts translunar coast, the Manned Space Flight Network for the first time will be called upon to track spacecraft position and to relay two-way communications, television and telemetry in a manned space flight to lunar distance.

Except for about 45 minutes or every two-hour lunar orbit, Apollo 8 will be "in view" of at least one of three 85-foot deep-space tracking antennas at Canberra, Australia, Madrid, Spain, and Goldstone, California.

Speculation arising from unmanned Lunar Orbiter missions was that mass concentrations below the lunar surface caused "wobbles" in the Spacecraft orbit. In Apollo 8 the ground network coupled with onboard navigational techniques will sharpen the accuracy of lunar orbit determination for future lunar missions.

Another facet of communicating with a manned spacecraft at lunar distance will be the use for the first time of the Apollo high-gain antenna — a four-dish unified S-band antenna that swings out from the service module after separation from the third stage.

The high-gain antenna relays onboard television and high bit-rate telemetry data, but should it become inoperative, the command module S-band omni antennas can relay voice communications, low bit-rate telemetry and spacecraft commands from the ground.

Apollo 8 will gather data on techniques for stabilizing spacecraft temperatures in deep-space operations by investigating the effects of rolling the spacecraft at a slow, fixed rate about its three axes to achieve thermal balance. The Apollo 8 mission will be the first opportunity for in-depth testing of these techniques in long periods of sunlight away from the reflective influence of the Earth.

Any solar flares occurring during the mission will be monitored by Solar Particle Alert Network (SPAN) stations around the world. Solar radiation and radiation in the Van Allen belt around the Earth present no hazard to the crew of Apollo 8 in the thick-skinned command module. The anticipated dosages are less than one rad per man, well below that of a thorough chest X-ray series.

Although Apollo 8's entry will be the first from a lunar flight, it will not be the first command module entry at lunar-return velocity.

The unmanned Apollo 4 mission in November 1967 provided a strenuous test of the spacecraft heat shield when the command module was driven back into the atmosphere from a 9,769 nautical mile apogee at 36,545 feet-per-second. By comparison, Apollo 8 entry velocity is expected to be 36,219 feet-per second. Heat shield maximum char depth on Apollo 4 was three-quarters of an inch, and heat loads were measured at 620 BTUs per square foot per second as compared to the 480 BTUs anticipated in a lunar-return entry.

Apollo 8 entry will be flown with a nominal entry range of 1,350 nautical miles in either the primary or backup control modes. Adverse weather in the primary recovery area can be avoided by a service propulsion system burn prior to one day before entry to shift the landing point. Less than one day out, the landing point can be shifted to avoid bad weather by using the spacecraft's 2,500 mile entry ranging capability.

The crew will wear the inflight coveralls during entry-pressure suits having been doffed and stowed since one hour after translunar injection. Experience in Apollo 7, when the crew flew the entry phase without pressure suit, helmets or gloves, prompted the decision not to wear suits once the spacecraft's pressure integrity was determined.

The decision to fly Apollo as a lunar orbit mission was made after thorough evaluation of spacecraft performance the ten-day Earth-orbital Apollo 7 mission in October and an assessment of risk factors involved in a lunar orbit mission. These risks are the total dependency upon the service propulsion engine for leaving lunar orbit and an Earth-return time as long as three days compared to one-half to three hours in Earth orbit.

Evaluated along with the risks of a lunar orbit mission was the value of the flight in furthering the Apollo program toward a manned lunar landing before the end of 1969. Principal gains from Apollo 8 will be

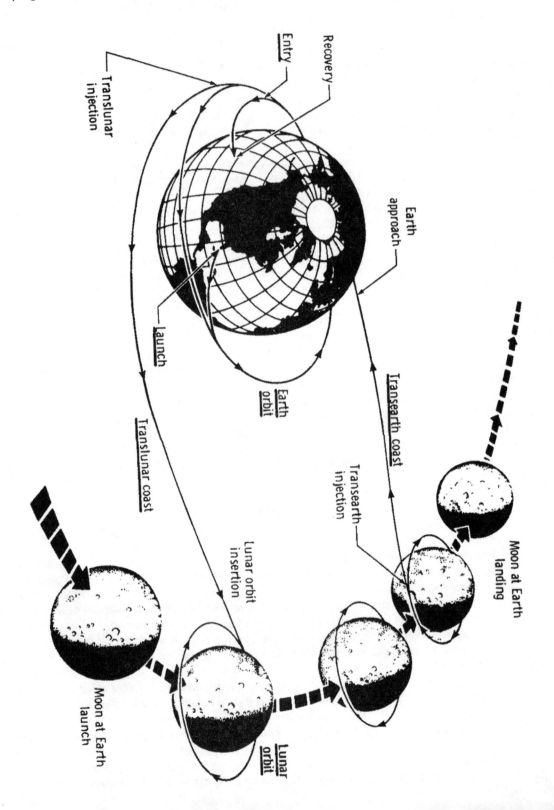

experience in deep space navigation, communications and tracking, greater knowledge of spacecraft thermal response to deep space, and crew operational experience — all directly applicable to lunar landing missions.

As many as seven live television transmissions may be made from Apollo 8 as it is on its path to the Moon, in orbit about the Moon and on the way back to Earth. The television signals will be received at ground stations and transmitted to the NASA Mission Control Center in Houston where they will be released live to commercial networks.

However, because of the great distances involved and the relatively low transmission power of the signals from the spacecraft to ground, the TV pictures are not expected to be of as high quality as the conventional commercial broadcast pictures.

Apollo 8 also will carry still and motion picture cameras and a variety or different films for black and white and color photography of the Moon and other items of interest. These include photographs of an Apollo landing site under lighting conditions similar to those during a lunar landing mission,

The Apollo 8 Saturn V launch vehicle is different from the two unmanned rockets that have preceded it in the following major aspects:

The updated J-2 engine capable of reaching a thrust of 230,000 pounds is being flown for the first time, on the third (S-IVB) stage.

The new helium prevalve cavity pressurization system will be flying on the first stage (S-IC) for the first time. In this system, four liquid oxygen prevalves have cavities filled with helium to create accumulators "shock absorbers" that will damp out the "pogo" effect.

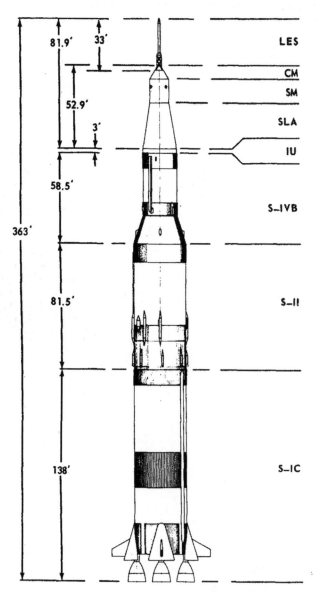

AS–503 CONFIGURATION
(APOLLO 8)

The liquid hydrogen engine feed line for each J-2 engine has been redesigned, and the auxiliary spark igniter lines have been replaced with lines without flex joints.

A helium heater will be used as a repressurization system on the S-IVB.

The center F-1 engine on the S-IC will be cut off early to keep the acceleration forces from building up past the four "G" level.

Software changes in the instrument unit will give a new cant capability to the outboard F-1 engines. After clearing the tower, the outboard engines will cant outward two degrees to reduce the load on the spacecraft in the event of a premature cutoff of an F-1 engine.

Film cameras will not be carried on the S-II stage to record the first and second plane separations.

The forward bulkhead of the S-II fuel tank is a lightweight type that will be used on future Saturn V vehicles.

(END OF GENERAL RELEASE; BACKGROUND INFORMATION FOLLOWS)

MISSION OBJECTIVES FOR APOLLO 8

* Demonstrate crew/space vehicle/mission support facilities performance during a manned Saturn V mission with command and service module.

* Demonstrate performance of nominal and selected mission activities including (1) translunar injection; (2) command service module navigation, communications and midcourse corrections; and (3) command service module consumables assessment and passive thermal control.

In addition, detailed test objectives have been designed to thoroughly wring out systems and procedures that have a direct bearing on future lunar landings and space operations in the vicinity of the Moon.

SEQUENCE OF EVENTS NOMINAL MISSION

Time From Lift-off (Hr:Min:Sec)	Event
00:00:00	Lift-off
00:01:17	Maximum dynamic pressure
00:02:06	S-IC Center Engine Cutoff
00:02:31	S-IC Outboard Engine Cutoff
00:02:32	S-IC/S-II Separation
00:02:33	S -II Ignition
00:02:55	Camera Capsule Ejection
00:03:07	Launch Escape Tower Jettison
	Mode I/ Mode II Abort Changeover
00:08:40	S-II Cutoff
00:08:41	S -II/S-IVB Separation
00:08:44	S-IVB Ignition
00:10:06	Mode IV Capability begins
00:10:18	Mode II/Mode III Abort Changeover
00:11:32	Insertion into Earth Parking Orbit
02:50:31	Translunar Injection Ignition
02:55:43	Translunar Injection Cutoff
	Trans lunar Coast Begins
03:09:14	S-IVB/CSM Separation
04:44:54	Begin Maneuver to Slingshot Attitude
05:07:54	LOX Dump Begins
05:12:54	LOX Dump Ends
TLI+ 6 Hrs.	Midcourse Correction 1
TLI+ 25 Hrs.	Midcourse Correction 2
LOI- 22 Hrs.	Midcourse Correction 3
LOI- 8 Hrs.	Midcourse Correction 4
69:07:29	Lunar Orbit Insertion (LOI-1) Initiation
69:11:35	Lunar Orbit Insertion (LOI-1) Termination
73:30:53	Lunar Orbit Insertion (LOI-2) Initiation
73:31:03	Lunar Orbit Insertion (LOI-2) Termination
89:15:07	Transearth Injection Initiate
89:18:33	Transearth Injection Terminate
TEI+15 Hrs.	Midcourse Correction 5
TEI+ 30 Hrs.	Midcourse Correction 6
TEI- 2 Hrs.	Midcourse Correction 7
146:49:00	Entry Interface
147:00:00	SPLASHDOWN

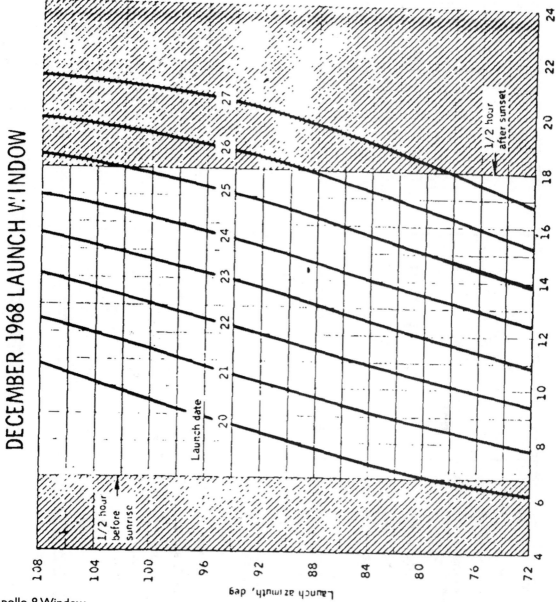

Apollo 8 Window

Pre - Launch

Apollo 8 is scheduled to be launched from Launch Complex 39, pad A, at Cape Kennedy, Florida on December 21, 1968. The launch window opens at 7:51 a.m. EST and closes at 12:32 p.m. EST. Should holds in the launch countdown or weather require a scrub, there are six days remaining in December during which the mission could be launched.

December Launch Days for Apollo 8	WINDOW (EST)	
	Open	Close
21	7:51 a.m.	12:32 p.m.
22	9:26 a.m.	2:05 p.m.
23	10:58 a.m.	3:35 p.m.
24	12:21 p.m.	4:58 p.m.
25	1:52 p.m.	6:20 p.m.
26	3:16 p.m.	6:20 p.m.
27	4:45 p.m.	6:20 p.m.

GO/NO-GO DECISION POINTS

MISSION PHASE	TIME OF DECISION	REMARKS
LAUNCH	REAL TIME	ORBIT IS "GO" IF Hp ≥ 75 NM
EARTH ORBIT	AFTER INSERTION	UNTIL LANDING AREA 2-1
	~ REV 1 U.S. PASS	TO TLI
	CRO	FOR TLI BURN
	~ TLI +10 -90 MINUTES	FOR ANY DISPERSIONS AND SEPARATION MANEUVER
	~ TLI + 2HR.	TO TLI +4HR.
	TLI + 4HR.	STILL PERFORM EARTH ORBITAL TYPE MISSION
TRANSLUNAR (CONTINUOUS MONITORING)	TLI + 5HR.	DEPENDENT UPON ΔV REQUIREMENTS FOR MID-COURSE CORRECTION: (1) CONTINUE MISSION (2) LUNAR FLYBY
	TLI + 6HR.	FOR MCC1
	ΔV < 4500 FPS TO PTP'S	FOR ANY MALFUNCTIONS THAT REQUIRE EARLY RETURNS
	TLI + 24HR.	FOR MCC2
	LOI - 20HR.	FOR MCC3
	LOI - 8HR.	FOR MCC4
LUNAR ORBIT	LOI_1 - 1HR.	FOR LOI BURN: AT LEAST 4HR. LUNAR ORBIT CAPABILITY.
	LOI_2 - 1HR.	FOR LOI CIRCULARIZATION BURN. 4HR ORBIT CAPABILITY.
	TEI - 1HR.	CONTINUOUS MONITORING WHILE IN VIEW FOR TRANS EARTH INJECTION BURN.

A variable launch azimuth of 72 degrees to 108 degrees capability will be available to assure a launch on time. This is the first Apollo mission which has employed the variable launch azimuth concept. The concept is necessary to compensate for the relative positioned relationship of the Earth at launch time.

MISSION DESCRIPTION

NOTE: Information presented in this press kit is based on a nominal mission. Plans may be altered prior to or during flight to meet changing conditions.

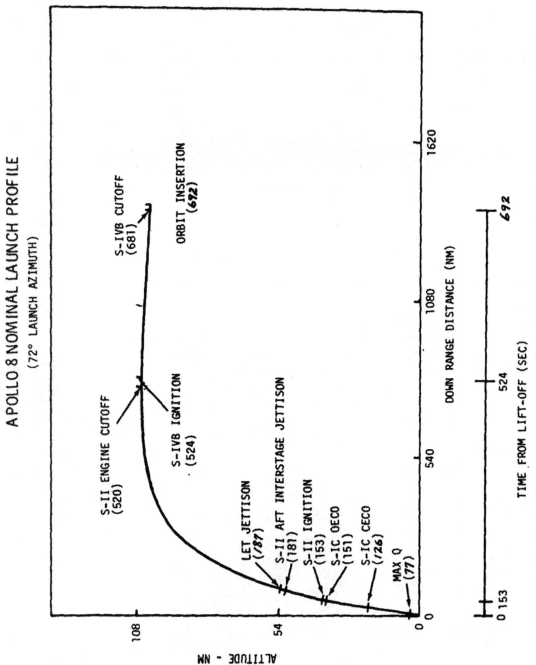

APOLLO 8 NOMINAL LAUNCH PROFILE
(72° LAUNCH AZIMUTH)

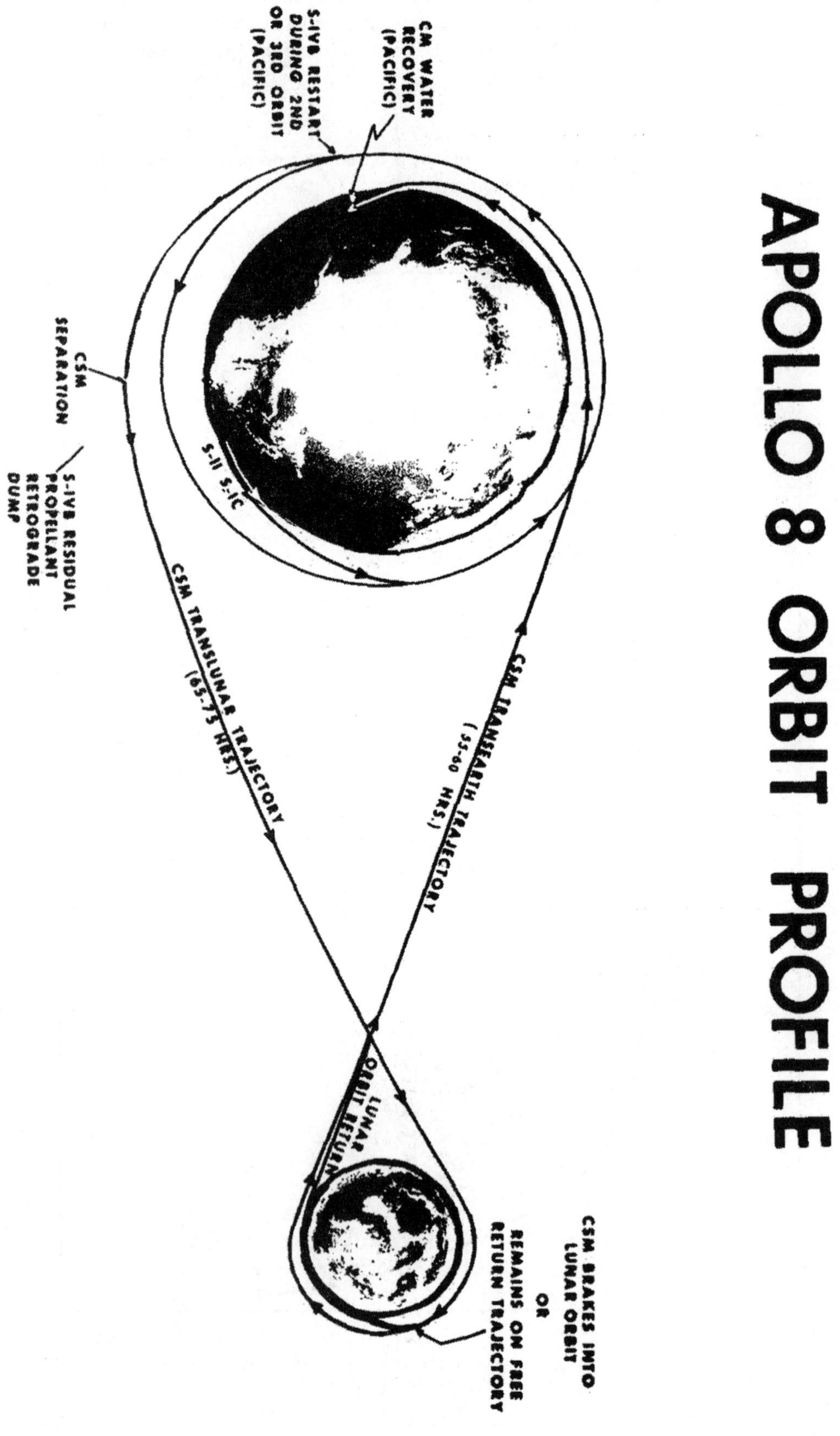

APOLLO 8 ORBIT PROFILE

CM WATER
RECOVERY
(PACIFIC)

S-IVB RESTART
DURING 2ND
OR 3RD ORBIT
(PACIFIC)

CSM
SEPARATION

S-IVB RESIDUAL
PROPELLANT
RETROGRADE
DUMP

S-II S-IC

CSM TRANSLUNAR TRAJECTORY
(65-75 HRS.)

CSM TRANSEARTH TRAJECTORY
(55-60 HRS.)

LUNAR
ORBIT RETURN

CSM BRAKES INTO
LUNAR ORBIT
OR
REMAINS ON FREE
RETURN TRAJECTORY

GO/NO-GO DECISION POINTS

MISSION PHASE	TIME OF DECISION	REMARKS
LAUNCH	REAL TIME	ORBIT IS "GC" IF $H_p \geq 75$ NM
EARTH ORBIT	AFTER INSERTION	UNTIL LANDING AREA 2-1
	~ REV 1 U.S. PASS	TO TLI
	CRO	FOR TLI BURN
TRANSLUNAR (CONTINUOUS MONITORING)	~ TLI +10 -90 MINUTES	FOR ANY DISPERSIONS AND SEPARATION MANEUVER
	~ TLI + 2HR.	TO TLI +4HR.
	TLI + 4HR.	STILL PERFORM EARTH ORBITAL TYPE MISSION
	TLI + 5HR.	DEPENDENT UPON ΔV REQUIREMENTS FOR MID-COURSE CORRECTION: (1) CONTINUE MISSION (2) LUNAR FLYBY
	TLI + 6HR.	FOR MCC_1
	$\Delta V < 4500$ FPS TO PTP'S	FOR ANY MALFUNCTIONS THAT REQUIRE EARLY RETURNS
	TLI + 24HR.	FOR MCC_2
	LOI - 20HR.	FOR MCC_3
	LOI - 8HR.	FOR MCC_4
LUNAR ORBIT	LOI_1 - 1HR.	FOR LOI BURN: AT LEAST 4HR. LUNAR ORBIT CAPABILITY.
	LOI_2 - 1HR.	FOR LOI CIRCULARIZATION BURN. 4HR ORBIT CAPABILITY.
	TEI - 1HR.	CONTINUOUS MONITORING WHILE IN VIEW FOR TRANS EARTH INJECTION BURN.

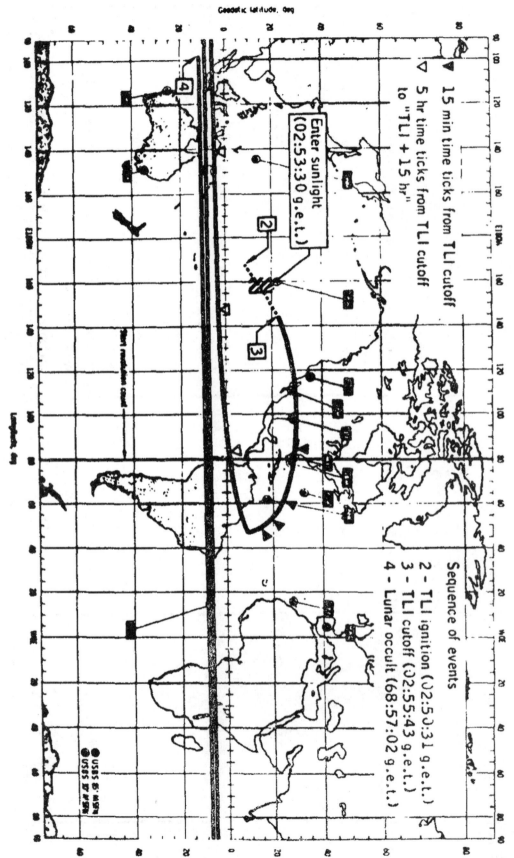

TRANSLUNAR COAST

▼ 15 min time ticks from TLI cutoff

▽ 5 hr time ticks from TLI cutoff
    to "TLI + 15 hr"

Enter sunlight
(02:53:30 g.e.t.)

Sequence of events

2 - TLI ignition (02:50:31 g.e.t.)
3 - TLI cutoff (02:55:43 g.e.t.)
4 - Lunar occult (68:57:02 g.e.t.)

LUNAR PARKING ORBIT

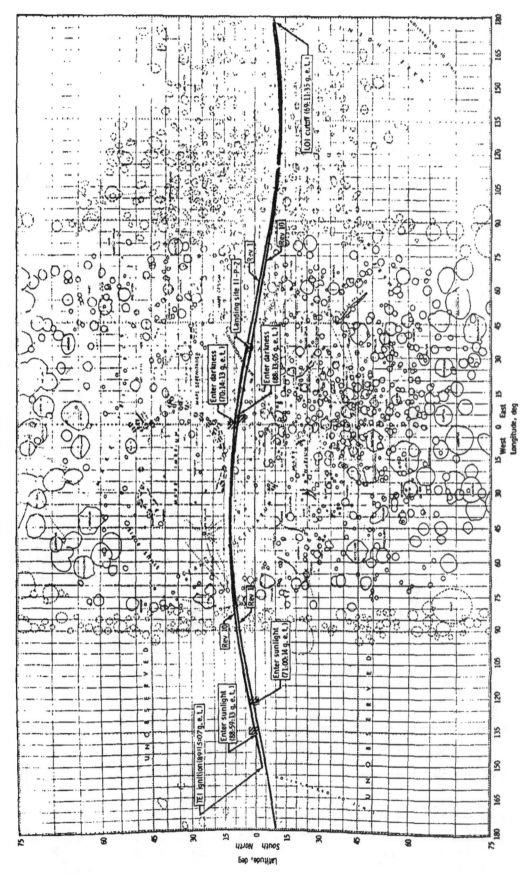

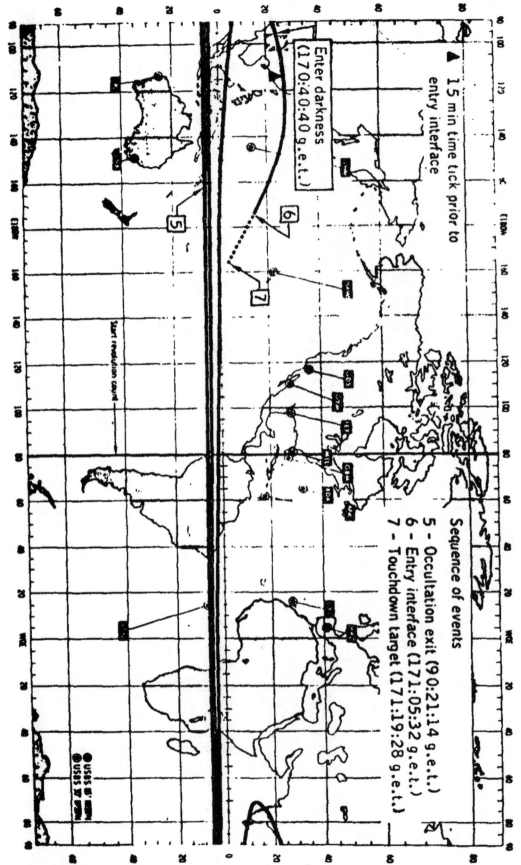

TRANSEARTH COAST

▲ 15 min time tick prior to entry interface

Enter darkness (170:40:40 g.e.t.)

5

6

7

Sequence of events

5 – Occultation exit (90:21:14 g.e.t.)
6 – Entry interface (171:05:32 g.e.t.)
7 – Touchdown target (171:19:28 g.e.t.)

Start revolution count

● USSS B'NER
◐ USSS IY FIBING

Launch Phase

Apollo 8 will be launched from Kennedy Space Center Launch Complex 39A on a launch azimuth that can vary from 72 degrees to 108 degrees, depending upon the time of day of launch. The azimuth changes with time of day to permit a fuel optimum injection from Earth parking orbit onto a free return circumlunar trajectory. Other factors influencing the launch windows are a daylight launch (sunrise -30 min. to sunset +30 min.), and proper sun angles on lunar landmarks in the Apollo landing zone.

The planned Apollo 8 launch date of December 21 will call for liftoff time at 7:51 a.m. EST on a launch azimuth of 72 degrees. Insertion into Earth parking orbit will occur at 11 min. 32 sec. ground elapsed time (GET) at an altitude of 103 nm (119 sm. 191.3 km). The orbit resulting from this launch azimuth will have an inclination of 32.5 degrees to the equator.

Earth Parking Orbit (EPO)

Apollo 8 will remain in Earth parking orbit after insertion and will hold a local horizontal attitude during the entire period. The crew will perform spacecraft systems checks in preparation for the Translunar Injection burn.

Translunar Injection (TLI)

In the second or third revolution in Earth parking orbit, the S-IVB third stage engine will reignite over the Pacific to inject Apollo 8 toward the Moon. The velocity will increase to 35,582 feet per second (10,900 meters/sec.). Injection will begin at an altitude of 106 nm (122 sm. 197 km) while the vehicle is in darkness. Midway through the translunar injection burn, Apollo 8 will enter sunlight.

Translunar Coast

Following the translunar injection burn, Apollo 8 will spend 66 hr. 11 min. in translunar coast. The spacecraft will separate from the S-IVB stage about 20 minutes after the start of the translunar injection burn, using the service module Reaction Control System (RCS) thrusters to maneuver out to about 50 to 70 feet from the stage for a 13-minute period of station keeping. The spacecraft will move away some five minutes later in an "evasive maneuver" while the S-IVB is commanded to dump residual liquid oxygen (LOX) through the J-2 engine bell about 1 hr. and 30 min. after separation. The third stage auxiliary propulsion system will be operated to depletion. The LOX dumping is expected to impart a velocity of about 90 fps to the S-IVB to lessen probability of recontact with the spacecraft; and place the stage in a "slingshot" trajectory passing behind the Moon's trailing edge and on into solar orbit.

Four midcourse correction burns are possible during the translunar coast phase, depending upon the accuracy of the trajectory. The first burn, at translunar injection +6 hrs., will be done if the needed velocity change is greater than 3 fps; the second at TLI +25 hrs.; the third at lunar orbit insertion (LOI) -22 hrs., and the fourth at LOI -8 hrs. The last three burns will be made only if the needed velocity is greater than 1 fps.

Lunar Orbit Insertion (LOI)

The first of two lunar orbit insertion burns will be made at 69:07:29 GET at an altitude above the Moon 69 nm (79 sm. 126.8 km). LOI No. 1 will have a nominal retrograde velocity change of 2,991 fps (912 m/sec) and will insert Apollo 8 into a 60 x 170 nm (69 x 196 sm. 111 x 314.8 km) elliptical lunar parking orbit. At 73:30:53, LOI Burn No. 2 will circularize the lunar parking orbit at 60 nm (69 sm. 111 km) with a retrograde velocity change or 138 fps(42.2 m/sec). The lunar parking orbit will have an inclination of 12 degrees to the lunar equator.

Lunar Parking Orbit (LPO)

During the 10 revolutions of lunar parking orbit, the Apollo 8 crew perform lunar landmark tracking and Apollo landing site tracking and photographic tasks, and stereo photography of the lunar surface from terminator to terminator. The last two lunar revolutions will be spent in preparation for the transearth injection burn.

APOLLO 8
LUNAR SEQUENCE OF EVENTS

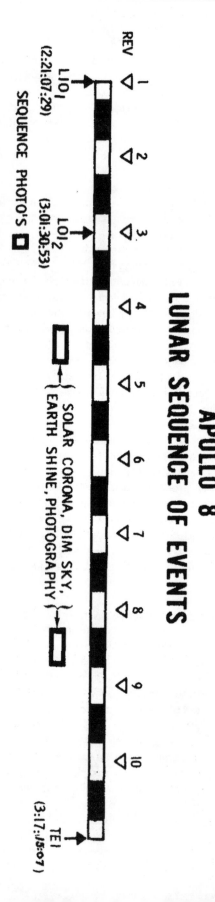

UMBRA* SCHEDULE		
REV	ENTER	EXIT
1	2:22:14:13	2:23:00:14
2	3:00:22:47	3:01:08:46
3	02:22:57	03:08:53
4	04:21:29	05:07:25
5	06:20:02	07:05:59
6	08:18:44	09:04:39
7	10:17:17	11:03:13
8	12:15:51	13:01:47
9	14:14:31	15:00:26
10	16:13:05	16:59:13

REV

△ 1   LIO₁ (2:21:07:29)

△ 2

△ 3   LOI₂ (3:01:30:53)

SEQUENCE PHOTO'S ▢

VERTICAL STEREO ▢ ▢

LIGHTING EVALUATION ▢ ▢

LANDMARK (LM) TRACKING
FOR DESCENT TARGETING ▮ I
UNKNOWN LM (ULM) TRACK :
(INCLUDES PHOTOGRAPHY) ▮

△ 4

△ 5   ▭ } SOLAR CORONA, DIM SKY,
      EARTH SHINE, PHOTOGRAPHY }→ ▭

△ 6

△ 7   ▮ ▮ ▮ ▮ ▮ ▮ ▮

OBLIQUE STEREO ▭

II ▮ IV ▮ 20° AFTER EXIT FROM DARK

III ▮ 30° BEFORE SUB-SOLAR

30° AFTER S-SOLAR

△ 8

△ 9

△ 10   TEI (3:17:u5:07)

* TIME IN GET BEGINNING AT END OF DAY 2. PENUMBRA
IS 13-15 SECONDS DURATION PRIOR TO AND AFTER UMBRA.

LUNAR ORBIT ACTIVITIES

Revolution	Activity
1	LOI-1
2	Housekeeping, systems checks, and landmarks sightings
3	LOI-2 , System checks and training photography
4,5,6,7	General landmark sightings, stereo photography, and general landmark photography
8	General photography, landmark sightings, Solar corona, Dim Sky, and Earth Shine photography
9	Prepare for TEI and perform oblique stereo strip photography
10	Perform TEI

Trans-Earth Injection (TEI)

The SPS trans-Earth Injection burn is nominally planned for 89:15:07 GET with a posigrade velocity increase of 3520 fps (1073 m/sec). The burn begins on the backside of the Moon and injects the spacecraft on a trajectory toward the Earth. It will reach 400,000 feet altitude above Earth at 146:49:00 GET.

Transearth Coast, Midcourse Maneuvers

During the approximate 57-hour Earth return trajectory, the Apollo 8 crew will perform navigation sightings on stars, and lunar and Earth landmarks, communications tests and spacecraft passive thermal control tests. Three midcourse corrections are possible during the transearth coast phase, and their values will be computed in real time. The midcourse corrections, if needed, will be made at transearth injection +15 hr., TEI +30 hr., and entry interface -2 hr. (400,000 ft. altitude).

Entry, Landing

Apollo 8 command module will be pyrotechnically separated from the service module approximately 15 minutes prior to reaching 400,000 ft. altitude. Entry will begin at 146:49:00 GET at a spacecraft velocity of 36,219 fps (11,005 m/sec). The crew will fly the entry phase with the G&N system to produce a constant deceleration (average 4 Gs) for a direct entry, rather than the dual-pulse "skip" entry technique considered earlier in Apollo program planning. Splashdown is targeted for the Pacific Ocean at 165 degrees West longitude by 4 degrees 55 min. North latitude. The landing footprint will extend some 1350 nm (1560 sm. 2497 km) from its entry point. Splashdown will be at 13 min. 46 sec. after entry.

CM END-OF-MISSION ENTRY AND LANDING POINTS*

Day Of Launch	Entry Point Latitude	Longitude	Landing Point Latitude	Longitude
21 Dec	14°42'N	174°30'E	4°55'N	165°00'W
22 Dec	5°35'N	173°50'E	1°00'S	165°00'W
23 Dec	1°20'N	174°35'E	8°10'S	165°00'W
24 Dec	10°15'S	172°15'E	12°50'S	165°00'W
25 Dec	18°55'S	171°25'E	18°00'S	165°00'W
26 Dec	25°00'S	170°45'E	22°10'S	165°00'W
27 Dec	22°25'S	170°55'E	25°25'S	165°00'W

*These points are for a 72 degree launch azimuth. Other launch azimuths will change the data slightly.

FLIGHT PLAN

Crew Activities

The Apollo 8 flight plan calls for at least one crewman to be awake at all times. The normal work/rest cycle will be 17 hours of work followed by seven hours of rest. The command module pilot and lunar module pilot sleep periods are scheduled simultaneously.

GEODETIC ALTITUDE VERSUS RANGE TO GO

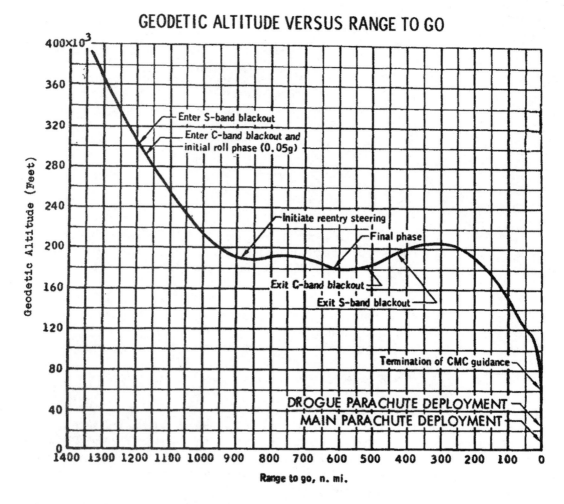

Following the translunar injection burn, the crew will take off their space suits and put on the inflight coveralls for the duration of the mission.

The Apollo 8 spacecraft normally will remain fully powered up throughout the entire mission, with the stabilization and control system and navigation sextant and scanning telescope turned on as needed. The inertial measuring unit and command module computer will stay in the "operate mode".

Changes of lithium hydroxide canisters for absorbing cabin carbon dioxide are scheduled at times when all crewmen are awake, and all scheduled maneuvers will be made when all crewmen are awake.

Flight plan updates will be relayed to the Apollo 8 crew over S-Band frequencies each day for the coming day's activities,

Following is a brief summary of tasks to be accomplished in Apollo 8 on a day-to-day schedule. The tasks are subject to changes to suit opportunity or operational factors:

Launch Day (0-24 hours):
* CSM systems checkout following Earth orbit insertion.
* Translunar Injection burn, Separation from S-IVB and transposition maneuver.
* Monitor S-IVB LOX blowdown and "slingshot" maneuver.
* Five sets of translunar coast star-Earth horizon navigation sightings.
* Perform first midcourse correction burn if required.
* Star-Earth landmark navigation sightings.

Second Day (24-48 hours):
* Midcourse correction burns Nos. 2 and 3 if required.
* Star-Earth horizon, star-lunar horizon navigation sightings.

Third Day (48-72 hours):
  * Star-lunar horizon navigation sightings.
  * Midcourse Correction burn No. 4 if required.
  * Lunar orbit insertion burn into initial 60 x 170 nm (69 x 196 sm. 111 x 314.8 km) orbit.
  * General lunar landmark observation, photography, onboard television.
  * Preparations for lunar orbit circularization burn.
  * Align IMU once during each lunar orbit dark period.

Fourth Day (72-96 hours):
  * Circularize lunar orbit to 60 nm (69 sm. 111 km).
  * Vertical stereo, convergent stereo navigation photography.
  * Solar corona photography.
  * Landmark and landing site tracking and photography.
  * Star-lunar landmark navigation sightings.
  * Transearth injection burn.

Fifth Day (96-120 hours):
  * Star-lunar horizon, star-Earth horizon navigation sightings.
  * Midcourse correction burns Nos. 5 and 6 if required.

Sixth Day (120-144 hours):
  * Star-Earth horizon and star-lunar horizon navigation sightings.
  * Midcourse correction burn No. 7 if required.

Seventh Day (144 hours to splashdown):
  * Star-Earth horizon and star-Earth landmark navigation sightings.
  * CM/SM separation, entry and splashdown.

Recovery Operations

The primary recovery line for Apollo 8 is in the mid-Pacific along the 165th west meridian of longitude where the primary recovery vessel, the aircraft carrier USS Yorktown will be on station. Nominal splashdown for a full duration lunar orbit mission launched on time December 21 will be at 4 degrees 55 minutes north x 165 degrees west at a ground elapsed time of 147 hours.

Other planned recovery lines for a deep-space mission are the East Pacific line extending parallel to the coastline of North and South America, the Atlantic Ocean line running along the 30th West meridian in the northern hemisphere and along the 25th West meridian in the southern hemisphere, the Indian Ocean line extending along the 65th East meridian, and the West Pacific line along the 150th East meridian in the northern hemisphere and jogging to the 170th East meridian in the southern hemisphere, Secondary landing areas for a possible Earth orbital alternate mission have been established in two zones in the Pacific and two in the Atlantic.

Ships on station in the launch abort area stretching 3,400 miles eastward from Cape Kennedy include the Helicopter landing platform USS Guadalcanal, one of whose duties will be retrieval of camera cassettes from the S-IC stage; the transport USS Rankin, the tracking ship USNS Vanguard which will be released from recovery duty after insertion into Earth parking orbit, and the oiler USS Chuckawan.

In addition to surface vessels deployed in the launch abort area and the primary recovery vessel in the Pacific, 16 HC-130 aircraft will be on standby at eight staging bases around the Earth: Tachikawa, Japan; Pago Pago, Samoa; Hawaii; Bermuda; Lajes, Azores; Ascension Island; Mauritius, and Panama Canal Zone.

Apollo 8 recovery operations will be directed from the Recovery Operations Control Room in the Mission Control Center and will be supported by the Atlantic Recovery Control Center, Norfolk, Va.; Pacific Recovery Control Center, Kunia, Hawaii; and control centers at Ramstein, Germany; and Albrook AFB, Canal Zone.

The Apollo 8 crew will be flown from the primary recovery vessel to Manned Spacecraft Center after recovery. The spacecraft will receive a preliminary examination, safing and power-down aboard the Yorktown prior to offloading at Ford Island, Hawaii, where the spacecraft will undergo a more complete

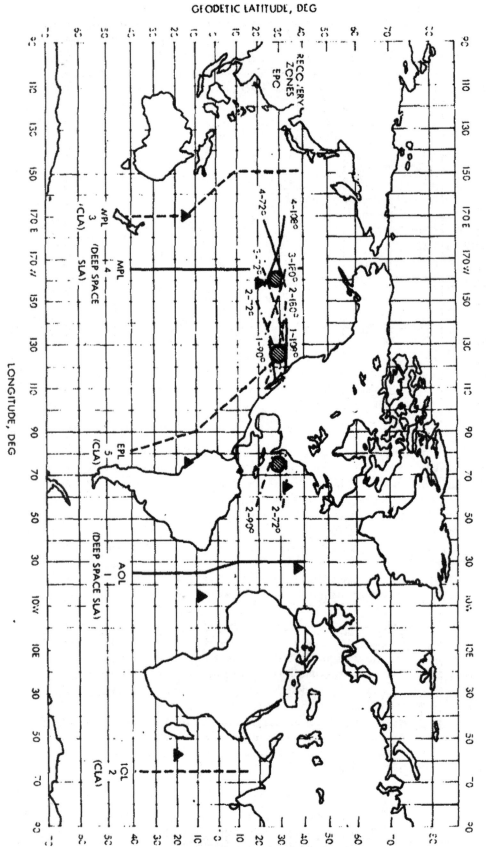

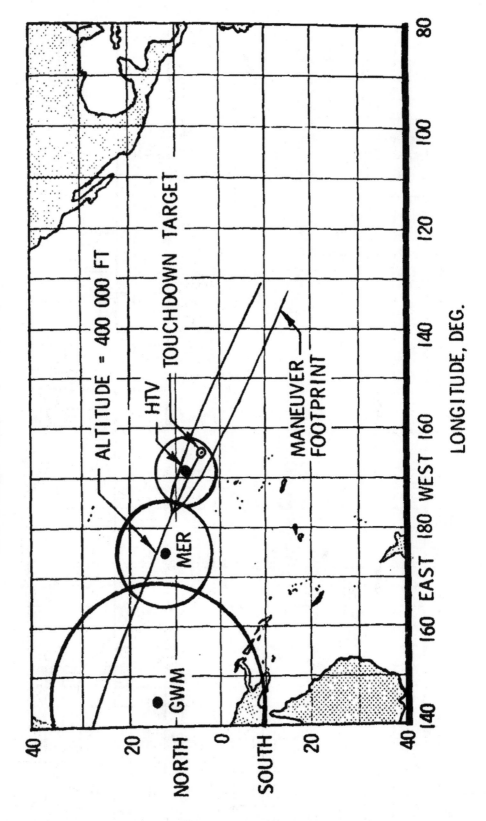

REENTRY COVERAGE
DECEMBER 21,1968-LAUNCH

APOLLO 8 PRIMARY LANDING AREA AND FORCE DEPLOYMENT

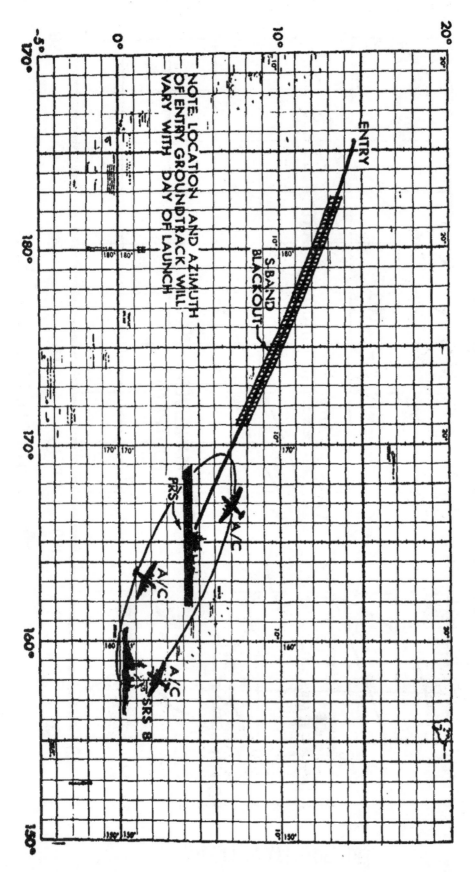

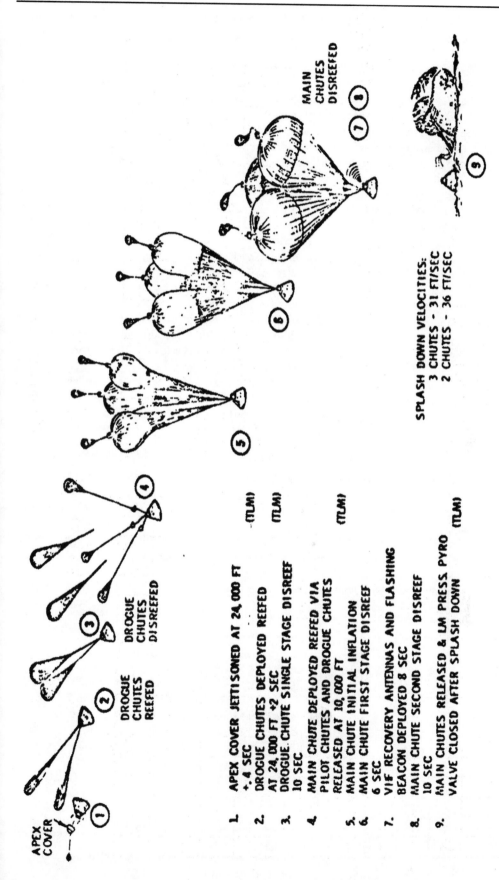

APEX COVER

1. APEX COVER JETTISONED AT 24,000 FT
   +.4 SEC   (TLM)

2. DROGUE CHUTES DEPLOYED REEFED
   AT 24,000 FT +2 SEC   (TLM)

3. DROGUE CHUTE SINGLE STAGE DISREEF
   10 SEC

4. MAIN CHUTE DEPLOYED REEFED VIA
   PILOT CHUTES AND DROGUE CHUTES
   RELEASED AT 10,000 FT   (TLM)

5. MAIN CHUTE INITIAL INFLATION

6. MAIN CHUTE FIRST STAGE DISREEF
   6 SEC

7. VHF RECOVERY ANTENNAS AND FLASHING
   BEACON DEPLOYED 8 SEC

8. MAIN CHUTE SECOND STAGE DISREEF
   10 SEC

9. MAIN CHUTES RELEASED & LM PRESS PYRO
   VALVE CLOSED AFTER SPLASH DOWN   (TLM)

DROGUE
CHUTES
REEFED

DROGUE
CHUTES
DISREEFED

MAIN
CHUTES
DISREEFED

SPLASH DOWN VELOCITIES:
   3 CHUTES - 31 FT/SEC
   2 CHUTES - 36 FT/SEC

Earth Landing System. Normal Sequence

deactivation. It is anticipated that the spacecraft will be flown from Ford Island to Long Beach, Calif., within 72 hours, and thence trucked to the North American Rockwell plant in Downey, Calif., for postflight analysis.

ALTERNATE MISSIONS

Several alternate mission plans have been prepared for the Apollo 8 mission, and the scope of the alternate is dependent upon when in the mission timeline it becomes necessary to switch to the alternate.

For example, if there is an early shutdown of the S-IVB stage during the Earth parking orbit insertion burn, the service propulsion system engine would be ignited for a contingency orbit insertion (COI), and again ignited later in the mission to boost the spacecraft to a 4,000 nm (4,610 sm. 7,400 km) apogee. The second burn into the high ellipse would be done only if the COI burn required less than 900 fps (274.5 m/sec) velocity increase. The COI-high apogee mission is Alternate 1, and would have a duration of up to 10 days.

Alternate mission 2 would be followed if the S-IVB failed to restart for the translunar injection burn out of Earth parking orbit. This alternate calls for the SPS engine to boost the spacecraft to the 4,000-nm (4,610 sm. 7,400 km) apogee for two to four revolutions. A deboost maneuver later would lower apogee and the mission would continue in low Earth orbit for 10 days.

Alternate mission 3 is split into three subalternates, each depending upon the apogee that can be reached after early S-IVB cutoff during the translunar injection burn. In the 3a alternate where apogee would be between 100 to 1,000 nm, (115 - 1,152 sm. 185 - 1,853 km) the orbit would be tuned up with an SPS burn to permit landmark sighting and the mission would follow the alternate 2 timeline. If apogee ranged between 1,000 nm (1,152 sm. 1853 km) and 25,000 nm (28,800 sm. 46,250 km), a phasing maneuver would be made at first perigee to shift a later perigee over a network station, where a deboost burn would lower apogee to 400 nm (461 sm. 740 km) and the mission would continue in low Earth orbit for the 3b alternate.

Alternate 3c would be followed if apogee was between 25,000 nm (28,800 sm. 46,250) and 60,000 nm (69,100 sm. 111,000 km). This alternate calls for a phasing maneuver at first perigee to shift later perigee to the recovery area. A second maneuver at the later perigee would adjust the elliptical orbit to one with a semi-synchronous period of about 12 hours, that is, there would be two daily perigee deorbit periods, one over the Pacific and one over the Atlantic. The entire mission would be flown in this type orbit, including direct entry from the high ellipse.

Early TLI cutoff which would produce an apogee greater than 60,000 nm (69,100 sm. 111,000 km) would fall into the alternate 3d category. This alternate calls for a circumlunar flyby using the Service Propulsion System to correct the flight profile back to a free-return trajectory. In some cases a lunar orbit mission may be possible.

All alternate mission plans call for water landings along the nominal Pacific recovery line or in the Atlantic and in general follow the lunar orbit mission timeline. Entry velocities from any of the alternates range between 26,000 fps and 36,000 fps (7,930-11,080 m/sec).

Apollo 8 alternate mission are summarized in the table below:

Apollo 8 Alternate Missions

Condition	Summary Alternate Plan
1. S-IVB early cutoff on EPO	If COI takes less than 900 fps burn, COI with SPS (274.5 m/sec), use SPS to raise apogee to 4,000 nm (4,610 sm. 7,200 km); If COI takes more than 900 fps (274.5 m/sec) remain in Earth orbit up to 10 days.
2. In EPO but S-IVB falls	Use SPS to raise apogee to 4,000 nm (4,610 sm. 7,400 km) to restart for TLI remain in high ellipse for 2-4 revolutions, then deboost to low Earth orbit for remainder of 10 days.
3. Early S-IVB TLI cutoff producing apogee of: a. 100-1,000 nm (115 - 1,152 sm.185 - 1,853 km)	Burn to 4,000 nm (4,610 sm.7,400 km) apogee, phase adjust for landmark sightings, remain in high ellipse for 2-4 revolutions, lower apogee and continue low Earth orbit mission.

b. 1,000-25,000 nm (1,152 - 28,800 sm. 1,853 - 46,250 km)	Make phasing maneuver at first perigee to shift later perigee over network station; at that perigee lower apogee to 400 nm (461 sm. 740 km); later SPS burn lowers apogee further and mission continues in low Earth orbit.
c. 25,000 - 60,000 nm (28,800 - 69,100 sm 46,250 - 111,000 kms)	Remain in established trajectory and make direct entry. (SPS fuel remaining not enough to lower apogee to 400 nm (461sm, 740 km) and still perform deorbit burn)
d. More than 60,000 nm (69,100 sm. 111,000 km)	Correct trajectory to lunar flyby and Earth free-return with SPS, direct entry.

ABORT MODES

The Apollo 8 mission can be aborted at anytime during the launch phase or during later phases after a successful insertion into earth orbit.

Abort modes can be summarized as follows:

Launch Phase

Mode I — Launch escape tower propels command module safely away from launch vehicle. This mode is in effect from about T-30 min. when LES is armed until LES jettison at 3:07 GET and command module landing point can range from the Launch Complex 39A area to 520 nm (600 sm. 964 km) downrange.

Mode II — Begins when LES is jettisoned and runs to 10:00 GET. Command module separates from launch vehicle and free-falls in a full-lift entry with landing between 400 and 3200 nm (461-3680 sm. 741-5930 km) downrange.

Mode III — Begins when full-lift landing point reaches 3200 nm (3680 sm. 5930 km) and extends through orbital insertion. The CSM would separate from the launch vehicle, and if necessary, an SPS retrograde burn would be made, and the command module would be flown half-lift to entry and landing between 3000 and 3350 nm (3450-3850 sm. 5560-6200 km) downrange.

Mode IV and Apogee Kick — Begins after the point the SPS could be used to insert the CSM into an earth parking orbit — from about 10 minutes after liftoff. The SPS burn into orbit would be made two minutes after separation from the S-IVB and the mission would continue as an earth orbit alternate, or if other conditions warranted, to landing in the West Atlantic or Central Pacific after one revolution, Mode IV is preferred over Mode III. A variation of Mode IV is the Apogee Kick in which the SPS would be ignited at first apogee to raise perigee and thereby set up a suitable orbit for a low earth-orbit alternate mission.

Earth Parking Orbit Phase

Aborts from earth parking orbit would be flown similar to the normal deorbit and entry that was flown on Apollo 7: SPS deorbit burn followed by CM/SM separation and guided entry.

Translunar Injection Phase

Aborts during the translunar injection phase are only a remote possibility, but if an abort became necessary during the TLI maneuver, an SPS retrograde burn could be made to produce spacecraft entry. This mode of abort would be used only in the event of an extreme emergency that affected crew safety. The spacecraft landing point would vary with launch azimuth and length of the TLI burn. Another TLI abort situation would be used if a malfunction cropped up after injection. A retrograde SPS burn at about 90 minutes after TLI shutoff would allow targeting to land in an Atlantic contingency landing area between 7½ and 17½ hours after initiating abort, depending on the change in velocity applied.

Translunar Coast Phase

Aborts arising during the three-day translunar coast phase would be similar in nature to the 90-minute TLI abort. Aborts from deep space bring into the play the moon's antipode (line projected from moon's center through earth's center to opposite face) and the effect of the earth's rotation upon the geographical location of the antipode. Abort times would be selected for landing when the antipode crosses 165 W Long. The antipode crosses the mid-Pacific recovery line once each 24 hours, and if a time-critical situation forces an abort earlier than the selected fixed abort times, landings would be targeted for the Atlantic Ocean, East Pacific, West Pacific or Indian Ocean recovery lines in that order of preference. From TLI plus 44 hours, a circumlunar abort becomes faster than an attempt to return directly to earth.

Lunar Orbit Insertion Phase

Early SPS shutdowns during the lunar orbit insertion burn (LOI) are covered by Modes I and III in the Apollo 8 mission (Mode II involves lunar module operations). Both modes would result in the CM landing the earth latitude of the moon antipode at the time the abort was performed. Mode I would be an SPS posigrade burn into an earth-return trajectory as soon as possible following LOI shutdown during the first two minutes of the LOI burn. Mode III occurs near pericynthion following one or more revolutions in lunar orbit. Following one or two lunar orbits, the Mode III posigrade SPS burn at pericynthion would inject the spacecraft into a transearth trajectory targeted for the mid-Pacific recovery line.

Lunar Orbit Phase

If during lunar parking orbit it became necessary to abort, the transearth injection (TEI) burn would be made early and would target Spacecraft landing to the mid-Pacific recovery line.

Transearth Injection Phase

Early shutdown of the TEI burn between ignition and two minutes would cause a Mode III abort and a SPS posigrade TEI burn would be made at a later pericynthion. Cutoffs after two minutes TEI burn time would call for a Mode I abort — restart of SPS as soon as possible for earth-return trajectory. Both modes produce mid-Pacific recovery line landings near the latitude of the antipode at the time of the TEI burn.

Transearth Coast Phase

Adjustments of the landing point are possible during the transearth coast through burns with the SPS or the service module RCS thrusters, but in general, these are covered in the discussion of transearth midcourse corrections. No abort burns will be made later than 20 hours prior to entry to avoid effects upon CM entry velocity and flight path angle.

PHOTOGRAPHIC TASKS

Photography seldom before has played as important a role in a space flight mission as it will on Apollo 8. The crew will have the task of photographing not only a lunar surface Apollo landing site to gather valuable data for subsequent lunar landing missions, but will also point their cameras toward visual phenomena in cislunar space which heretofore have posed unanswered questions.

A large quantity of film of various types has been loaded aboard the Apollo 8 spacecraft for lunar surface photography and for items of interest that crop up in the course of the mission.

Camera equipment carried on Apollo 8 consists of two 70mm Hasselblad still cameras with two 80mm focal length lenses a 250mm telephoto lens, and associated equipment such as filters, ringsight, spotmeter and intervalometer for stereo strip photography. For motion pictures a 16mm Maurer data acquisition camera with variable frame speed selection will be used. Accessories for the motion picture camera include lenses of 200, 75, 18 and 5mm focal lengths, a right-angle mirror, a command module boresight bracket and a power cable.

Photographic tasks have been divided into three general categories: lunar stereo strip photography, engineering photography and items of interest.

Apollo 8 photographic tasks are summarized as follows:

<u>Lunar Stereo Strip Photography</u> — overlapping stereo 70mm frames shot along the lunar orbit ground track with spacecraft aligned to local vertical. Photos will be used for terrain analysts and photometric investigations.

<u>Engineering Photography</u> — Through-the-window photography of immediate region around spacecraft to gather data on existence of contaminant cloud around the spacecraft and to further understand source of window visibility degradation. Cabin interior photography documenting crew activities will also be taken as an aid to following flight crews.

<u>Items Of Interest</u> — Dim-light targets: Gegenschein (a round or elongated spot of light in space at a point 180° from the sun) photos on one-minute exposure with spacecraft held in inertial attitude on dark side of Moon and during translunar and transearth coast; zodiacal light along the plane of the ecliptic (path of Sun around celestial sphere), one-minute exposures during dark side of lunar orbit; star fields under various lighting conditions to study effect of spacecraft debris clouds and window contamination on ability to photograph stars; lunar surface in earthshine to gain photometric data about lunar surface under low-level illumination.

<u>Lunar surface in daylight at zero phase angle</u> (spacecraft shadow directly below spacecraft) to further measure reflective properties of lunar surface; lunar terminator in daylight at oblique angles to evaluate capability of terrain analysis on such photos; Apollo exploration sites, Surveyor landing sites and specific features and areas to augment present lunar surface photography and to correlate with Surveyor photos; image motion compensation with long-length lenses by tracking target with spacecraft; phenomena, features and other items of interest selected by the crew in real time: and lunar seas through red and blue filters for correlating with color discontinuities observed from earth.

Apollo 8 film stowage is as follows: 3 magazines of Panatomic-X intermediate speed black and white for total 600 frames; 2 magazines SO-368 Ektachrome color reversal for total 352 frames; 1 magazine SO-121 Ektachrome special daylight color reversal for total 160 frames; and 1 magazine 2485 high-speed black and white (ASA 6,000, push to 16,000) for dim-light photography, total 120 frames. Motion picture film: nine 130-foot magazines SO-368 for total 1170 feet, and two magazines SO-168 high speed interior color for total 260 feet.

TELEVISION

As many as seven live television transmissions are being considered during the Apollo 8 flight. Up to three transmissions are being considered on the way to the Moon, one to two from lunar orbit, and possibly two on the way back from the Moon.

The frequency and duration of TV transmissions are dependent upon the level of various other mission activities and the availability of the spacecraft's high gain antenna. This antenna is primarily used to transmit engineering data, which has priority over TV transmissions.

The TV signals will be sent from the spacecraft to ground stations at Goldstone, California and Madrid, Spain, where the signal will be converted to commercial frequencies. The TV signal will be released live to public networks from the Mission Control Center, Houston.

Because of crew activities from launch through translunar injection, TV operations are not planned prior to the translunar coast phase of the mission about 12 hours into the flight.

The purpose of television during Apollo 8 is to evaluate TV transmission at lunar distances for planning future lunar missions and to provide live TV coverage of the Apollo 8 flight to the public.

The 4.5 pound RCA TV camera is equipped with 160 degree and 9 degree field of view lens. A 12-foot power-video cable permits the camera to be hand-held at the command module windows for the planned photography.

In TV broadcast to homes, the average distance is only five miles between the transmitting station and the home TV set, while the station transmits an average of 50,000 watts of power. In contrast, the Apollo 8 TV camera operates on only 20 watts, and on this mission, will be over 200,000 miles from the home TV sets.

The NASA ground station's large antenna and sensitive receivers make up for most of this difference, but the Apollo pictures are not expected to be as high quality as normal broadcast programs.

SPACECRAFT STRUCTURE SYSTEMS

Apollo spacecraft No. 103 for the Apollo 8 mission is comprised of a launch escape system, command module, service module and a spacecraft-lunar module adapter. The latter serves as a mating structure to the instrument unit atop the S-IVB stage of the Saturn V for this mission, lunar module test article B (LTA-B) will be housed in the adapter.

Launch Escape System/(LES) — Propels command module to safety in an aborted launch. It is made up of an open frame tower structure mounted to the command module by four frangible bolts, and three solid-propellant rocket motors: a 155,000-pound-thrust launch escape system motor, a 3,000 pound-thrust pitch control motor that bends the command module trajectory away from the launch vehicle and pad area. Two canard vanes near the top deploy to turn the command module aerodynamically to an attitude with the heat-shield forward. Attached to the base of the Escape System is a boost protective cover composed of glass, cloth and honeycomb, that protects the command module from rocket exhaust gases from the main and the jettison motor. The system is 33 feet tall, four feet in diameter at the base and weighs 8,900 pounds (4040 kg).

Command Module(CM) Structure — The basic structure of the Command Module is a pressure vessel encaged in heat-shields, cone-shaped 12 feet high, base diameter of 12 feet 10 Inches, and launch weight 12,392 pounds (5626 kg).

The command module consists of the forward compartment which contains two negative pitch reaction control engines and components of the Earth landing system; the Crew Compartment, or inner pressure vessel, containing crew accommodations, controls and displays, and spacecraft systems; and the aft compartment housing ten reaction control engines and fuel tankage.

Heat-shields around the three compartments are made of brazed stainless steel honeycomb with an outer layer of phenolic epoxy resin as an ablative material. Heat-shield thickness, varying according to heat loads, ranges from 0.7 inches (at the apex) to 2.7 inches on the aft side.

The spacecraft inner structure is of aluminum alloy sheet-aluminum honeycomb bonded sandwich ranging in thickness from 0.25 inches thick at forward access tunnel to 1.5 inches thick at base.

Service Module (SM) Structure — The service module is a cylinder 12 feet 10 inches in diameter by 22 feet long. For the Apollo 8 mission, it will weigh 51,258 pounds (23,271 kg) at launch. Aluminum honeycomb panels one inch thick form the outer skin, and milled aluminum radial beams separate the interior into six sections containing service propulsion system and reaction control fuel-oxidizer tankage, fuel cells and onboard consumables.

Spacecraft-LM Adapter (SLA) Structure — The spacecraft LM adapter is a truncated cone 20 feet long tapering from 260 inches diameter at the base to 154 inches at the forward end at the service module mating line. Aluminum honeycomb 1.75 inches thick is the stressed-skin structure for the spacecraft adapter. The SLA weighs 4,150 pounds (1,884 kg).

SPACECRAFT SYSTEMS

Guidance, Navigation and Control System/(GNCS) — Measures and controls spacecraft attitude and velocity, calculates trajectory, controls spacecraft propulsion system thrust vector and displays abort data. The Guidance System consists of three subsystems: inertial, made up of inertial measuring unit and associated power and data components; computer, consisting of display and keyboard panels and digital computer which processes information to or from other components; and optic, including scanning telescope, sextant for celestial and/or landmark spacecraft navigation.

Stabilization and Control System/(SCS) — Controls spacecraft rotation, translation and thrust vector and provides displays for crew-initiated maneuvers; backs up the guidance system. It has three subsystems; attitude reference, attitude control and thrust vector control.

Service Propulsion System/(SPS) — Provides thrust for large spacecraft velocity changes and de-orbit burn through a gimbal-mounted 20,500-pound-thrust hypergolic engine using nitrogen tetroxide oxidizer and a 50-50 mixture of unsymmetrical dimethyl hydrazine and hydrazine fuel. Tankage of this system is in the service module. The system responds to automatic firing commands from the guidance and navigation system or to manual commands from the crew. The engine provides a constant thrust rate. The stabilization and control system gimbals the engine to fire through the spacecraft center of gravity.

Reaction Control System (RCS) — This includes two independent systems for the command module and the service module. The service module reaction controls have four identical quads of four 100-pound thrust hypergolic engines mounted, near the top of the Service Module, 90 degrees apart to provide redundant spacecraft attitude control through cross-coupling logic inputs from the Stabilization and Guidance Systems. Small velocity change maneuvers can also be made with the Service Module reaction controls. The Command Module Reaction Control System consists of two independent six-engine subsystems of 94 pounds thrust each. One is activated after separation from the Service Module, and is used for spacecraft attitude control during entry. The other is maintained in a sealed condition as a backup. Propellants for both systems are monomethyl hydrazine fuel and nitrogen tetroxide oxidizer with helium pressurization. These propellants are hypergolic, i.e. they burn spontaneously on contact without need for an igniter.

Electrical Power System/(EPS) — Consists of three, 31 cell Bacon-type hydrogen-oxygen fuel cell power plants in the Service Module which supply 28-volt DC power, three 28-volt DC zinc-silver oxide main storage batteries in the Command Module lower equipment bay, two pyrotechnic batteries in the Command Module lower equipment bay, and three 115-200-volt 400-cycle three-phase AC inverters powered by the main 28-volt DC bus. The inverters are also located in the lower equipment bay. Supercritical cryogenic hydrogen and oxygen react in the fuel cell stacks to provide electrical power, potable water and heat. The Command Module main batteries can be switched to fire pyrotechnics in an emergency. A battery charger builds the batteries to full strength as required.

Environmental Control System/(ECS) — Controls spacecraft atmosphere, pressure and temperature and manages water. In addition to regulating cabin and suit gas pressure, temperature and humidity, the system removes carbon dioxide, odors and particles, and ventilates the cabin after landing. It collects and stores fuel cell potable water for crew use, supplies water to the glycol evaporators for cooling, and dumps surplus water overboard through the urine dump valve. Excess heat generated by spacecraft equipment and crew is routed by this system to the cabin heat exchangers, to the space radiators, to the glycol evaporators, or it vents the heat to space.

Telecommunication System — Consists of pulse code modulated telemetry for relaying to Manned Space Flight Network stations data on spacecraft systems and crew condition, VHF/AM and unified S-Band tracking transponder, air-to-ground voice communications, onboard television, and a VHF recovery beacon. Network stations can transmit to the spacecraft such items as updates to the Apollo guidance computer and central timing equipment, and real-time commands for certain onboard functions.
         The Apollo high-gain steerable S-Band Antenna will be flown for the first time on the Apollo 8 mission. Deployed shortly after CSM separation from the S-IVB stage, the high-gain antenna will be tested in the two-way mode between the spacecraft and the Manned Space Flight Network stations during translunar coast, lunar orbit and Earth return.
         The high-gain S-Band Antenna consists of four, 31-inch diameter parabolic dishes mounted on a folding boom at the aft end of the service module. Nested alongside the service propulsion system engine nozzle until deployment, the antenna swings out at right angles to the spacecraft longitudinal axis, with the boom pointing 52 degrees below the heads-up horizontal. Signals from the ground stations can be tracked either automatically or manually with the antenna's gimbaling system. All normal S-Band voice and uplink /downlink communications will be handled by the high-gain antenna.

Sequential System — Interfaces with other spacecraft systems and subsystems to initiate critical functions during launch, docking maneuvers, pre-orbital aborts and entry portions of a mission. The system also controls routine spacecraft sequencing such as Service Module separation and deployment of the Earth landing system.

Emergency Detection System/(EDS) — Detects and displays to the crew launch vehicle emergency conditions, such as excessive pitch rates or two engines out, and automatically or manually shuts down the booster and activates the launch escape system; functions until the spacecraft is in orbit.

Earth Landing System/(ELS) — Includes the drogue and main parachute system as well as post-landing recovery aids. In a normal entry descent, the Command Module apex cover is jettisoned at 24,000 feet, followed by two mortar-deployed reefed 16.5-foot diameter drogue parachutes for orienting and decelerating the spacecraft. After drogue release, three pilot chutes pull out the three main 83.3-foot diameter parachutes with two-stage reefing to provide gradual inflation in three steps. Two main parachutes out of three will provide a safe landing.

Recovery aids include the uprighting system, swimmer interphone connections, sea dye marker, flashing beacon, VHF recovery beacon and VHF transceiver. The uprighting system consists of three compressor-inflated bags to turn the spacecraft upright if it should land in the water apex down (Stable II position).

Caution and Warning System — Monitors spacecraft systems for out-of-tolerance conditions and alerts crew by visual and audible alarms so that crewmen may trouble-shoot the problem.

Controls and Display

Provide readouts and control functions of all other spacecraft systems in the command and service modules. All controls are designed to be operated by crewmen in pressurized suits. Displays are grouped according to the frequency the crew refers to them.

SATURN V LAUNCH VEHICLE

The white Saturn V launch vehicle, with the Apollo spacecraft and launch escape system mounted atop, towers 363 feet above the launch pad. The three propulsive stages and the instrument unit have a combined height of 281 feet. The vehicle weighs 6,219,760 pounds at ignition.

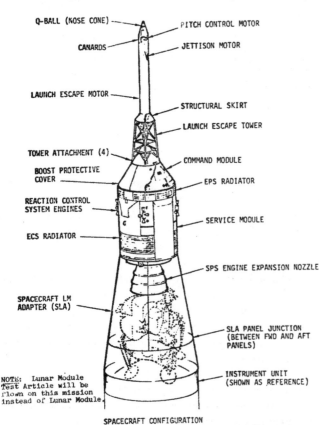

SPACECRAFT CONFIGURATION

Marked with black paint in sections for better optical tracking, identified with huge red lettering, and wearing the United States Flag on the first stage, the giant vehicle is capable of hurling 285,000 pounds into low Earth orbit or sending about 100,000 pounds to the Moon.

First Stage

The first stage (S-IC) of the Saturn V is 138 feet tall and 33 feet in diameter, not including the fins and engine shrouds on the thrust structure. It was developed jointly by the National Aeronautics and Space Administration's Marshall Space Flight Center, Huntsville, Ala., and The Boeing Co.

Marshall assembled four S-IC stages: a structural test model, a static test version and the first two flight stages. The first flight stage launched Apollo 4 on the first Saturn V flight Nov. 9, 1967. The second S-IC launched Apollo 6 on April 4, 1968.

Boeing, as prime contractor, built two ground test units. Boeing is responsible for assembly of the other 13 flight stages at Marshall's Michoud Assembly Facility in

New Orleans. The first flight model S-IC built by Boeing is the first stage of the AS-503 launch vehicle.

The static test model and the first three flight versions were fired at the Marshall Space Flight Center Test Laboratory. All other S-IC stages are being test fired at Marshall's Mississippi Test Facility in Hancock County, Miss.

Dry weight of the first stage is 305,650 pounds. It's two propellant tanks have a total capacity of 4.4 million pounds of fuel and oxidizer — some 202,000 gallons (1,352,711 pounds) of RP-1 (kerosene) and 329,000 gallons (3,130,553 pounds) of liquid oxygen. Stage weight at separation, including residual propellants, will be 380,738 pounds. The normal propellant flow rate to the five F-1 engines is 28,000 pounds per second. The five engines produce a combined thrust roughly equivalent to 180 million horsepower at maximum speed.

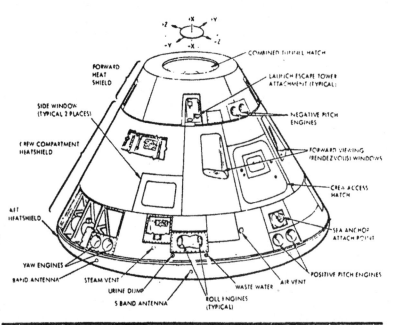

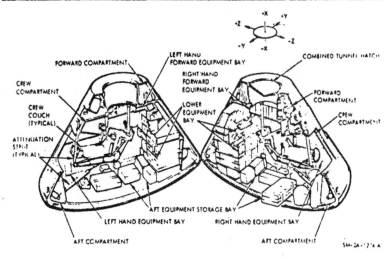

First Stage

During the planned 151 seconds of burn time, the engines will propel the Apollo/Saturn V to an altitude of 36.3 nautical miles (41.9 sm. 67 km) and carry it downrange 47.4 nautical miles (54.6 sm. 88 km), making good a speed of 5,267,3 knots (6,068 statute miles-per-hour) at first stage engine cutoff.

Four of the engines are mounted on a ring, each 90 degrees from its neighbor. These four can be gimbaled to control the rocket's direction of flight. The fifth engine is mounted rigidly in the center.

Second Stage

The second stage (SII) is 81.5 feet tall and 33 feet in diameter. It weighs 88,600 pounds dry, 1,035,463 pounds loaded with propellant. Weight at separation will be 103,374 pounds.

The 14,774 pounds difference between dry weight and weight at separation includes the 12,610 pound, S-IC/S-II interstage section, 2,164 pounds of ullage rocket propellants and other items on board.

The stage's two propellant tanks carry about 271,800 gallons (152,638 pounds) of liquid hydrogen and 87,500 gallons (792,714 pounds) of liquid oxygen. Its five J-2 engines develop a combined thrust of 1 million pounds.

The second stage carries the rocket to an altitude of 105.8 nautical miles (121.9 sm. 197 km) and a distance of some 805 nautical miles (927.4 sm. 1490 km) downrange. Before burnout it will be moving 13,245 knots (15,258.3 mph). The J-2 engines will run six minutes and seven seconds.

The Space Division of North American Rockwell Corp., builds the second stage at Seal Beach,

California. The cylindrical vehicle is made up of the forward skirt (to which the third stage connects), the liquid hydrogen tank, the liquid oxygen tank, the thrust structure (on which the engines are mounted and an interstage section to which the first stage connects. The tanks are separated by an insulated common bulkhead.

North American Rockwell conducted research and development static testing at the Santa Susana, Calif., test facility and at the NASA-Mississippi Test Facility. The flight stage for the Apollo 8 was shipped via the Panama Canal for captive firings at Mississippi Test Facility.

Third Stage

The third stage (S-IVB) was developed by the McDonnell Douglas Astronautics Co. at Huntington Beach, Calif. It is the larger and more powerful successor to the S-IV that served as the second stage of the Saturn I. The third stage is flown from its manufacturing site to the McDonnell Douglas' Test Center, Sacramento, Calif., for static test firings. The stage is then flown to the NASA Kennedy Space Center.

Measuring 58 feet 5 inches long by 21 feet 8 inches in diameter, the stage weighs 26,000 pounds dry. At separation in flight its weight will be 29,754 pounds exclusive of the liquid hydrogen and liquid oxygen in the main tanks. This extra weight consists mainly of solid and liquid propellants used in retro and ullage rockets and in the auxiliary propulsion system (APS).

An interstage section connects the second and third stages. This 8,760-pound section stays with the second stage at separation, exposing the single J-2 engine mounted on the thrust structure. The after skirts, connected to the interstage at the separation plane, encloses the liquid oxygen tank which holds some 20,400 gallons of the oxidizer. Above this is the large fuel tank holding about 77,200 gallons of liquid hydrogen. Weight of the S-IVB and payload at insertion into parking orbit will be 283,213 pounds. Weight at injection into translunar trajectory will be 122,380 pounds.

Total usable propellants carried in the two tanks 18,234,509 pounds, with fuel and oxidizer separated by an insulated common bulkhead. Insulation is necessary in both upper stages because liquid oxygen, at about 293 degrees below zero F. is too warm for liquid hydrogen, at minus 423 degrees.

The aft skirt also serves as a mount for two auxiliary propulsion system modules spaced 180

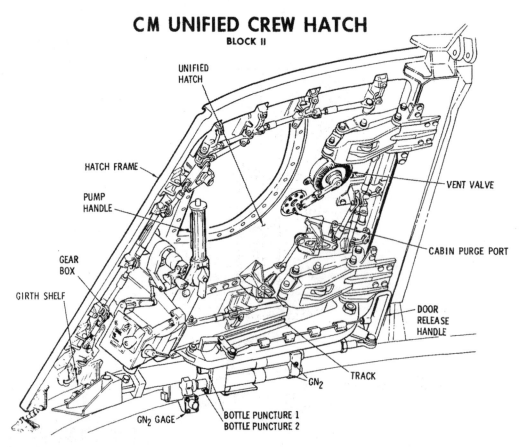

CM UNIFIED CREW HATCH
BLOCK II

SERVICE MODULE
BLOCK II

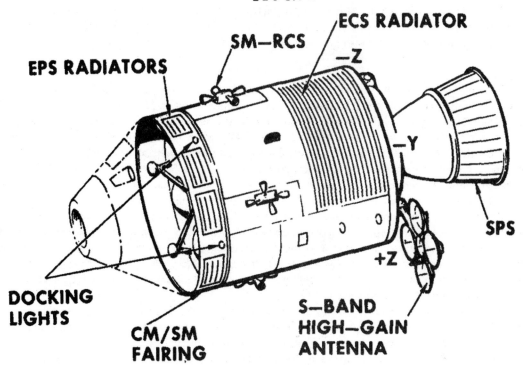

EPS RADIATORS

SM–RCS

ECS RADIATOR

–Z

–Y

SPS

+Z

S–BAND HIGH–GAIN ANTENNA

DOCKING LIGHTS

CM/SM FAIRING

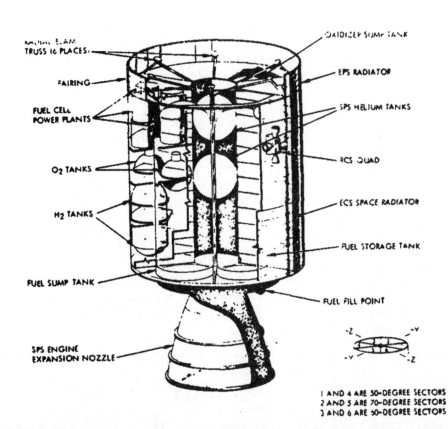

MAIN BEAM TRUSS 16 PLACES

FAIRING

FUEL CELL POWER PLANTS

O₂ TANKS

H₂ TANKS

FUEL SUMP TANK

SPS ENGINE EXPANSION NOZZLE

OXIDIZER SUMP TANK

EPS RADIATOR

SPS HELIUM TANKS

RCS QUAD

ECS SPACE RADIATOR

FUEL STORAGE TANK

FUEL FILL POINT

1 AND 4 ARE 50-DEGREE SECTORS
2 AND 5 ARE 70-DEGREE SECTORS
3 AND 6 ARE 50-DEGREE SECTORS

degrees apart. Each module contains three liquid-fueled 147-pound thrust engines, one each for roll, pitch and yaw, and a 72-pound-thrust, liquid-fueled ullage engine.

Four solid-propellant retro-rockets of 37,500 pounds thrust each are mounted on the interstage to back the second stage away from the third stage at separation. The third stage also carries two solid-propellant ullage motors of 3,400 pounds thrust each. These motors help to move the third stage forward and away from the second stage upon separation and serve the additional purpose of settling the liquid propellants in the bottoms of the tanks in preparation for J-2 ignition. The first J-2 burn is 152 seconds, the second, 5 min. 12 sec.

Propulsion

The 41-rocket engines of the Saturn V have thrust ratings ranging from 72 pounds to more than 1.5 million pounds. Some engines burn liquid propellants, others use solids.

The five F-1 engines in the first stage burn RP-1 (kerosene) and liquid oxygen. Each engine in the first stage develops 1.415 million pounds of thrust at liftoff, building up to 1.7 million pounds thrust before cutoff. The cluster of five F-1's gives the first stage a thrust range from 7.57 million pounds at liftoff to 8.5 million pounds just before cutoff.

The F-1 engine weighs almost 10 tons, is more than 18 feet high and has a nozzle-exit diameter of nearly 14 feet. The F-1 undergoes static testing for an average 650 seconds in qualifying for the 150-second run during the Saturn V first stage booster phase. This run period, 800 seconds, is still far less than the 2,200 seconds of the engine guarantee period. The engine consumes almost three tons of propellants per second.

The first stage of the Saturn V for this mission has four other rocket motors. These are the solid-fuel retrorockets which will slow and separate the stage from the second stage. Each rocket produces a thrust of 87,900 pounds for 0.6 second.

The main propulsion for the second stage is a cluster of five J-2 engines burning liquid hydrogen and liquid oxygen. Each engine develops a mean thrust of 200,000 pounds (variable from 175,000 to 225,000 in phases or flight), giving the stage a total mean thrust of 1 million pounds.

Designed to operate in the hard vacuum of space, the 3,500-pound J-2 is more efficient than the F-1 because it burns the high-energy fuel hydrogen.

The second stage also has four 21,000-pound-thrust solid fuel rocket engines. These are the ullage rockets mounted on the interstage section. These rockets fire to settle liquid propellant in the bottom of the main tanks and help attain a "clean" separation from the first stage, then they drop away with the interstage at second plane separation.

Fifteen rocket engines perform various functions on the third stage. A single J-2 provides the main propulsive force; there are two main ullage rockets, four retro-rockets and eight smaller engines in the auxiliary propulsion system.

Instrument Unit

The Instrument Unit (IU) is a cylinder three feet high and 21 feet 8 inches in diameter. It weighs 4,880 pounds.

Components making up the "brain" of the Saturn V are mounted on cooling panels fastened to the inside surface of the instrument unit skin. The refrigerated "cold plates" are part of a system that removes heat by circulating fluid coolant through a heat exchanger that evaporates water from a separate supply into the vacuum of space.

The six major systems of the instrument unit are structural, thermal control, guidance and control, measuring and telemetry, radio frequency and electrical.

The instrument unit maintains navigation, guidance and control of the vehicle; measurement of vehicle performance and environment; data transmission with ground stations; radio tracking of the vehicle; checkout and monitoring of vehicle functions; detection of emergency situations; generation and network distribution of electric power for system operation; and preflight checkout and launch and flight operations.

A path-adaptive guidance scheme is used in the Saturn V instrument unit. A programmed trajectory is used in the initial launch phase with guidance beginning only after the vehicle has left the atmosphere. This is to prevent movements that might cause the vehicle to break apart while attempting to compensate for winds, jet streams and gusts encountered in the atmosphere.

If such air currents displace the vehicle from the optimum trajectory in climb, the vehicle derives a new trajectory. Calculations are made about once each second throughout the flight. The launch vehicle digital computer and launch vehicle data adapter perform the navigation and guidance computations.

The ST-124M inertial platform — the heart of the navigation, guidance and control system — provides space-fixed reference coordinates and measures acceleration along the three mutually perpendicular axes of the coordinate system.

International Business Machines Corp. is prime contractor for the instrument unit and is the supplier of the guidance signal processor and guidance computer. Major suppliers of instrument unit components are: Electronic Communications, Inc., control computer; Bendix Corp., ST-124M inertial platform; and IBM Federal Systems Division, launch vehicle digital computer and launch vehicle data adapter.

Launch Vehicle Camera Systems

Fewer cameras will be carried aboard the Saturn V launch vehicle on the Apollo 8 mission than on the previous Saturn V flights. The first stage will carry four motion picture cameras and two television cameras. The film cameras will be ejected for recovery.

First Stage — Film Cameras

The four film cameras will be mounted on the inside of the forward skirt of the first stage. Two cameras (1 and 3) will be mounted lens forward and canted inward five degrees to view the separation of the first and second stages. These two cameras will start 144 seconds after liftoff and run about 40 seconds. Two cameras (2 and 4) will be mounted lens-aft, with lenses connected by fiber optic bundles to manhole covers in the top of the liquid oxygen tank. These covers provide viewing windows and mounts for strobe lights. The tank will be lighted inside by pulsed strobe to enable the cameras to record the behavior of the liquid oxygen in flight. These two cameras will be turned on 30 seconds before liftoff; the strobe lights will be turned off shortly before first stage cutoff.

The film cameras, loaded with color film, are carried in recoverable capsules inserted in ejection tubes. They will be ejected 177 seconds after vehicle liftoff, or 25 seconds after stage separation, at 49 nautical miles (56.4 sm. 90.8 km) altitude at a point 76.8 nautical miles (88.5 sm. 142 km) downrange. Camera impact is expected some 413.2 nm (473 sm. 763 km) downrange about 11 minutes after liftoff.

First Stage — Television

Both television cameras will be mounted inside the thrust structure of the first stage.

A fiber optic bundle from each camera will split into two separate bundles going to lenses mounted outside the heat shield in the engine area. This will provide two images for each camera, or four for the system. The images will be tilted 90 degrees from vertical to give a wider view as two images appear on each cathode ray tube. Each lens will view the center engine and one outer engine, thus providing a view of each outer engine working in conjunction with the fixed center engine.

A removable aperture disk can be changed from f/2 to f/22 to vary the image intensity. (Fiber optics reduce image intensity about 70 per cent.) A quartz window, rotated by a DC motor with a friction drive, protects the objective lens on the end of the fiber optic bundle. Fixed metallic mesh scrapers will remove soot from the rotating windows. Images from the two objective lenses are combined into the dual image in the larger fiber optics bundle by a "T" fitting. A 14-element coupling lens adapts the large dual image bundle to the camera.

Images from both cameras are multiplexed together for transmission on a single telemetry link. The images are unscrambled at the receiving station.

The video system cameras contain 28-volt vidicon cameras, pre-amplifiers and vertical sweep circuits for 30 frames per second scanning.

The TV cameras operate continuously from 55 minutes before liftoff until destroyed on first stage reentry.

Ejection and Recovery

When the camera capsules are ejected, stabilization flaps open for the initial part of the descent. When the capsules descend to about 15,000 feet above ground a para-balloon will inflate automatically, causing flaps to fall away. A recovery radio transmitter and flashing light beacon are turned on about 6 seconds after para-balloon inflation. After touchdown, the capsule effuses a dye marker to aid sighting, and it releases a shark repellent to protect the capsules, para-balloon (which keeps the capsule afloat) and the recovery team.

A Navy ship with helicopters and frogmen aboard will be cruising in the splashdown area for the camera capsules. The capsules will be picked up and flown by helicopter to Kennedy Space Center. The capsules will then be transferred to a data plane at Patrick Air Force Base, Fla., for immediate transfer to the Marshall Center at Huntsville where the film will be processed.

Sequence of Events

NOTE: Information presented in this press kit is based on a nominal mission. Plans may be altered prior to or during flight to meet changing conditions.

Launch

The first stage of the Saturn V will carry the vehicle and Apollo spacecraft to an altitude of 36.3 nautical miles (41.9 sm. 67 km) and 47.4 nautical miles (54.6 sm. 88 km) downrange, building up speed to 5,267.3 knots (6,068 mph) in two minutes 31 seconds of powered flight. After separation from the second stage, the first stage will continue a ballistic trajectory ending in the Atlantic Ocean some 357.6 nautical miles (412 sm. 665 km) downrange from Cape Kennedy (latitude 30.24 degrees N and longitude 74.016 degrees W) about nine minutes after liftoff.

Second Stage

The second stage, with engines running 6 minutes and 7 seconds will propel the vehicle to an altitude of about 105.8 nautical miles (121.9 sm. 197 km) some 805 nautical miles (927.4 sm. 1490 km) downrange, building up to 13,245 knots (15,258.3 mph) space fixed velocity. The spent second stage will land in the Atlantic Ocean about 19 minutes after lift-off some 2,190 nautical miles (2,527 sm. 407 km) from the launch site, at latitude 31.79 degrees N and longitude 38.24 degrees W.

First Third-Stage Burn

The third stage, in its 152-second initial burn, will place itself and the Apollo spacecraft in a circular orbit 103 nautical miles (119 sm. 191.3 km) above the Earth. Its inclination will be 32.5 degrees and orbital period, 88.2 minutes. Apollo 8 will enter orbit at about 47.08 degrees W longitude and 26.33 degrees N latitude at a velocity of 25,592 feet-per-second (17,433 statute mph or 15,132 knots).

Parking orbit

While in the two revolutions in Earth parking orbit, the Saturn V third stage and spacecraft systems will be checked out in preparation for the second S-IVB burn.

Second Third-Stage Burn

Near the end of the second revolution, the J-2 engine of the third stage will be reignited for 5 minutes 12 seconds. This will inject the vehicle and spacecraft into a translunar trajectory. About 20 minutes later the CSM separates from the S-IVB/IU. Following separation, the S-IVB performs an attitude maneuver in preparation for dumping LOX residuals and a burn to depletion of the S-IVB auxiliary propulsion system (APS). Dumping of S-IVB LOX residuals and APS Burn on Apollo lunar missions may be done to alter the velocity and trajectory of the spent S-IVB-IU to place it in a "slingshot" trajectory passing behind the Moon's trailing edge into solar orbit.

Differences In Apollo 6 and Apollo 8 Launch Vehicles

The new helium prevalve cavity pressurization system will be flying on the S-IC for the first time. In this system, cavities in the liquid oxygen prevalves are filled with helium to create accumulators or "shock absorbers" to damp out oscillations. This system was installed to prevent excessive longitudinal oscillations experienced in the Apollo 6 flight.

The center engine of the S-IC stage will be cut off early (126 seconds after liftoff) in the boost phase. This is being done to keep acceleration forces from passing the four "g" level.

Software changes in the instrument unit will give a new cant capability to the outboard F-1 engines of the first stage. The engines will be in the normal position until the vehicle clears the launch tower. The engines will then cant outward two degrees to reduce the load on the spacecraft in the event one of the F-1 engines cut off prematurely.

The S-IC stage will also be carrying added instrumentation on this flight. The added instruments will monitor the actions of the first stage in connection with the accumulators installed to reduce longitudinal oscillations.

On the S-II stage, the propellant utilization system will have an open loop capability for improved reliability. Commands from the instrument unit will keep the propellants close to the planned levels.

The liquid hydrogen engine feed line for each J-2 engine has been redesigned, and the auxiliary spark igniter lines have been replaced.

Leaks developed in the Augmented Spark Igniter (ASI) lines of two J-2 engines on the second Saturn V launched. The new lines, without flex joints, were tested on the Apollo 7 flight and verified as satisfactory.

Film cameras will not be carried on the second stage of Apollo 8. On Apollo 6 two recoverable cameras recorded the first and second plane separations of the first and second stages.

A lightweight forward bulkhead in the liquid hydrogen tank is being used in this vehicle. The original bulkhead was damaged during production of the S-II stage about two years ago. The new bulkhead is the same type which will be used in S-II stages of all future Saturn V vehicles.

Added instrumentation will also be flown on the S-II stage. The added equipment will monitor the actions of the new ASI lines and further verify their suitability for future flights.

The thrust of the J-2 engines on this S-II stage will reach almost 230,000 pounds at one point. (Average engine thrust will be 228,290.7 pounds for a total stage thrust of 1,141,453.7 pounds.) The engines will ignite at a mixture ratio of 5.0 to 1 and then shift to 5.5 to 1 for the first portion of the burn. The high thrust occurs during the high mixture ratio burn phase. The mixture ratio then shifts at 440 seconds after liftoff back to 4.5 to 1 for the remainder of the powered flight. (Average engine thrust is 180,168.7 pounds for a stage total thrust of 900, 843.7 pounds.)

The propellant utilization system for the S-IVB stage will also have the open loop capability.

The J-2 engine of the third stage is an uprated version with a top thrust of 230,000 pounds at the mixture ratio of 5.5 to 1. However, the engine will not reach 230,000 pounds thrust on this flight because the 5.5 to 1 mixture ratio is not planned. It is the first uprated J-2 engine to be flown.

A helium heater will be used as a repressurization system on the S-IVB for the first time on this flight. In this system, cold helium from the storage bottles are heated to provide pressurization for the fuel tank.

The capability to reignite the S-IVB stage engine after separation from the spacecraft has been added to this vehicle. The prime mission plan calls for the stage to be reignited in Earth parking orbit with the spacecraft attached. The added capability will permit stage reignition even if some unforeseen trouble requires separation of the spacecraft from the stage while in Earth parking orbit.

For example, loss of an F-1 engine during S-IC powered flight might require the use of too much S-IVB propellant in reaching Earth parking orbit. This would leave a shortage of propellant for the translunar injection burn. Also, a condition might be discovered during parking orbit which would pose a risk to the astronauts if the S-IVB was restarted while the spacecraft was attached. In any event of separation, the S-IVB will still be restarted, if possible, so that test data can be obtained from the stage even though it will be unmanned.

The redesigned ASI line will also be flown on this flight. It will be monitored again to verify satisfactory findings on the AS-205 flight. Instrumentation for checking out the new line has also been added, while research and development instrumentation no longer needed in some other parts of the vehicle has been removed.

Launch Vehicle Instrumentation and Communications

A total of 2,608 measurements will be taken in flight on the Saturn V launch vehicle. This includes 893 in the first stage, 978 in the second stage, 359 in the third stage and 378 in the instrument unit.

The Saturn V will carry 20 telemetry systems, six on the first stage, six on the second stage, three on the third stage and five on the instrument unit. The vehicle will carry a radar tracking system on the first stage and a C-Band system and command system on the instrument unit. Each propulsive stage will carry a range safety system as on previous flights. Four motion picture cameras in recoverable capsules and two television cameras will also be on the first stage.

APOLLO 8 LAUNCH OPERATIONS

NASA's John F. Kennedy Space Center is responsible for preflight checkout, test and launch of the Apollo 8 space vehicle. A government-industry team of about 450 will conduct the final countdown from

Firing Room 1 of the Launch Control Center (LCC).

The firing room team is backed up by more than 5,000 persons who are directly involved in launch operations at KSC — from the time the vehicle and spacecraft stages arrive at the center until the launch is completed.

Two major decisions made while flight hardware for Apollo 8 was at Kennedy Space Center had a pronounced affect on the scheduling of checkout and launch operations at KSC.

The first was the decision in April of this year to fly Apollo 8 as the first manned Saturn V flight. At that time, the launch vehicle had been erected on its mobile launcher in high bay number 1 of the Vehicle Assembly Building (VAB).

As a result of this decision, it was necessary to disassemble the stages so that the second stage (S-II) could be returned to Marshall Space Flight Center's Mississippi Test Facility (MTF). There it underwent modifications to prepare it for manned flight and to ensure that the problem which caused premature shutdown of two engines on the previous flight was corrected.

While the second stage was at MTF, modification kits were sent to KSC for installation on the first (S-IC) and third (S-IVB) stages to correct the launch vehicle longitudinal vibration anomaly and the failure of the third-stage engine to reignite in orbit. Both of these problems occurred on the second unmanned Saturn V flight, Apollo 6.

The second stage was returned to KSC in late June and erected on the booster in high bay number 1 in July. The third stage and instrument unit were erected the following month.

In August the decision was made not to fly a lunar module on Apollo 8. Lunar Module 3, which had been scheduled for the mission, had been at KSC since June, and was carried as part of the Apollo 8 spacecraft checkout schedule. As a result of this decision, LM-3 was dropped from the spacecraft schedule and a lunar module test article was inserted.

Assembly and checkout of Apollo 8 has been carried out while launch teams at KSC prepared and launched the Apollo 7 mission and began preparation for the Apollo 9 and Apollo 10 missions, scheduled for the first and second quarter of next year.

On Oct. 9, when the Apollo 8 space vehicle was moved from the Vehicle Assembly Building to Pad A, the launch team for Apollo 7 was conducting the final countdown to launch for the first manned Apollo mission. The Apollo 9 launch vehicle was already erected in the Vehicle Assembly Building and its Command, Service, and Lunar modules were undergoing checkout in the Manned Spacecraft Operations Building (MSOB) in the industrial area. Lunar Module 4, the first flight hardware for Apollo 10 to arrive at KSC, was also undergoing checkout in the MSOB.

The Apollo 8 command and service modules arrived at KSC in mid-August and were taken to the Manned Spacecraft Operations Building for checkout and altitude chamber runs. The prime and backup crews each participated in altitude chamber tests during which spacecraft and crew systems were checked out at simulated altitudes in excess of 200,000 feet.

At the completion of testing in the altitude chamber, the CSM was mated to the spacecraft lunar module adapter (SLA) and moved to the VAB where it was mechanically mated to the launch vehicle. The move to the VAB and erection on the launch vehicle was completed Oct 7. Following installation of certain ordnance items and the launch escape system, the space vehicle was moved to Pad A by the crawler transporter.

Integrated systems testing was conducted at the pad and the space vehicle was electrically mated in early November

The first overall test of the space vehicle, called the Overall Test, Plugs In, verified the compatibility of the space vehicle systems, ground support equipment, and off-site support facilities by demonstrating the ability of the systems to proceed through a simulated countdown, launch, and flight. During the simulated flight portion of the test, the systems were required to respond to both emergency and normal flight conditions.

The space vehicle Flight Readiness Test was conducted in mid-November. This was the last overall test before the countdown demonstration. Both the prime and backup crews participate in portions of the FRT, which is a final overall test of the vehicle systems and associated ground support equipment when all systems are as near as possible to a launch configuration.

After hypergolic fuels were loaded aboard the space vehicle and RP-1, the launch vehicle first stage fuel was brought aboard, and the final major test of the space vehicle began. This was the countdown demonstration test (CDDT), a dress rehearsal for the final countdown to launch. The CDDT for Apollo 8 was divided into a "wet" and a "dry" portion. During the first, or "wet" portion, the entire countdown, including propellant loading, was carried out down to T-8.9 seconds. The astronaut crews did not participate in the wet CDDT. At the completion of the wet CDDT, the cryogenic propellants (liquid oxygen and liquid hydrogen) were off-loaded, and the final portion of the countdown was re-run, this time simulating the fueling and with the prime astronaut crew participating as they will on launch day.

Because of the complexity involved in the checkout of the 363-foot-tall Apollo/Saturn V configuration, the launch teams make use of extensive automation in their checkout. Automation is one of the major differences in checkout used on Apollo compared to the procedures used in the Mercury and Gemini programs.

RCA 110A computers, data display equipment and digital data techniques are used throughout the automatic checkout from the time the launch vehicle is erected in the VAB through liftoff. A similar, but separate computer operation called ACE (Acceptance Checkout Equipment) is used to verify the flight readiness of the spacecraft. Spacecraft checkout is controlled from separate firing rooms located in the Manned Spacecraft Operations Building.

Countdown

Hrs : Min: Secs:

T		
	- 28: 00: 00	Start launch vehicle and spacecraft countdown activities
	- 24: 30: 00	S-II power-up
	- 24: 00: 00	S-IVB power-up
	- 09: 00: 00	Six-hour built-in hold
	- 09: 00: 00	End of built-in hold close CM & boost protective cover hatch
	- 08: 59: 00	Clear pad for launch vehicle cryo loading
	- 07: 28: 00	Start S-IVB LOX loading
	- 07: 04: 00	S-IVB LOX loading complete; start S-II LOX loading
	- 06: 27: 00	S-II LOX loading complete; start S-IC LOX loading
	- 04: 57: 00	S-IC LOX loading complete
	- 04: 54: 00	Start S-II liquid hydrogen loading
	- 04: 11: 00	S-II liquid hydrogen loading complete; start S-IVB liquid hydrogen loading
	- 03: 30: 00	Flight crew departs Manned Spacecraft Operations building
	- 03: 28: 00	S-IVB liquid hydrogen loading Complete
	- 03: 13: 00	Closeout crew on station; start ingress preps
	- 02: 40: 00	Start flight crew ingress
	- 02: 10: 00	Flight crew ingress complete
	- 01: 40: 00	Close spacecraft hatch
	- 01: 00: 00	Start RP-1 level adjust
	- 00: 42: 00	Arm Launch Escape System pyro buses
	- 00: 35: 00	RP-1 level adjust complete
	- 00: 15: 00	Spacecraft on internal power
	- 00: 05: 30	Arm Safe and Arm devices
	- 00: 03: 07	Terminal Count Sequence (TCS) start
	- 00: 00: 17.2	Guidance reference release command
	- 00: 00: 08.9	S-IC ignition command
	- 00: 00: 00	Lift-off

NOTE: The foregoing is the Apollo 8 countdown that was prepared as the press kit was ready for printing. Some changes may be made as a result of the Countdown Demonstration Test (CDDT).

KSC Launch Complex 39

Launch Complex 39 facilities at the Kennedy Space Center were planned and built specifically for the Saturn V program, the space vehicle that will be used to carry astronauts to the Moon.

Complex 39 introduced the mobile concept of launch operations, a departure from the fixed launch pad techniques used previously at Cape Kennedy and other launch sites. Since the early 1950's when the first ballistic missiles were launched, the fixed launch concept had been used on NASA missions. This method called for assembly, checkout and launch of a rocket at one site — the launch pad. In addition to tying up the pad, this method also often left the flight equipment exposed to the outside influences of the weather for extended periods.

Using the mobile concept, the space vehicle is thoroughly checked in an enclosed building before it is moved to the launch pad for final preparations. This affords greater protection, a more systematic checkout process using computer techniques, and a high launch rate for the future, since the pad time is minimal.

Saturn V stages are shipped to the Kennedy Space Center by ocean-going vessels and specially designed aircraft, such as the Guppy. Apollo spacecraft modules are transported by air. The spacecraft components are first taken to the Manned Spacecraft Operations Building for preliminary checkout. The Saturn V stages are brought immediately to the Vehicle Assembly Building after arrival at the nearby turning basin.

Apollo 8 is the third Saturn V to be launched from Pad A, Complex 39. The historic first launch of the Saturn V, designated Apollo 4, took place Nov. 9, 1967 after a perfect countdown and on-time liftoff at 7 a.m. EST. The second Saturn V mission — Apollo 6 — was conducted last April 4.

The major components of Complex 39 Include: (1) the Vehicle Assembly Building (VAB) where the Apollo 8 was assembled and prepared; 2) the Launch Control Center, where the launch team conducts the preliminary checkout and countdown; (3) the mobile launcher, upon which the Apollo 8 was erected for checkout and from where it will be launched; (4) the mobile service structure, which provides external access to the space vehicle at the pad; (5) the transporter, which carries the space vehicle and mobile launcher, as well as the mobile service structure to the pad; (6) the crawlerway over which the space vehicle travels from the VAB to the launch pad; and (7) the launch pad itself.

The Vehicle Assembly Building

The Vehicle Assembly Building is the heart of Launch Complex 39. Covering eight acres, it is where the 363-foot tall space vehicle is assembled and tested. The VAB contains 129,482,000 cubic feet of space. It is 716 feet long, and 518 feet wide and it covers 343,500 square feet of floor space. The foundation of the VAB rests on 4,225 steel pilings, each 16 inches in diameter, driven from 150 to 170 feet to bedrock. If placed end to end, these piles would extend a distance of 123 miles. The skeletal structure of the building contains approximately 60,000 tons of structural steel. The exterior is covered by more than a million square feet of insulated aluminum siding.

The building is divided into a high bay area 525 feet high and a low bay area 210 feet high, with both areas serviced by a transfer aisle for movement of vehicle stages.

The low bay work area, approximately 442 feet wide and 274 feet long, contains eight stage-preparation and checkout cells. These cells are equipped with systems to simulate stage interface and operation with other stages and the instrument unit of the Saturn V launch vehicle.

After the Apollo 8 launch vehicle upper stages arrived at the Kennedy Space Center, they were moved to the low bay of the VAB. Here, the second and third stages underwent acceptance and checkout testing prior to mating with the S-IC first stage atop mobile launcher No. 1 in the high bay area.

The high bay provides the facilities for assembly and checkout of both the launch vehicle and spacecraft. It contains four separate bays for vertical assembly and checkout. At present, three bays are equipped, and the fourth will be reserved for possible changes in vehicle configuration.

Work platforms — some as high as three-story buildings — in the high bays provide access by surrounding the launch vehicle at varying levels. Each high bay has five platforms. Each platform consists of two bi-parting sections that move in from opposite sides and mate, providing a 360-degree access to the section of the space vehicle being checked.

A 10,000-ton-capacity air conditioning system, sufficient to cool about 3,000 homes, helps to control the environment within the entire Office, laboratory, and workshop complex located inside the low bay area of the VAB. Air conditioning is also fed to individual platform levels located around the vehicle.

There are 141 lifting devices in the VAB, ranging from one-ton hoists to two 250-ton high-lift bridge cranes.

The mobile launchers, carried by transporter vehicles, move in and out of the VAB through four doors in the high bay area, one in each of the bays. Each door is shaped like an inverted T. They are 152 feet wide and 114 feet high at the base, narrowing to 76 feet in width. Total door height is 456 feet.

The lower section of each door is of the aircraft hangar type that slides horizontally on tracks.

Above this are seven telescoping vertical lift panels stacked one above the other, each 50 feet high and driven by an individual motor. Each panel slides over the next to create an opening large enough to permit passage of the Mobile Launcher.

The Launch Control Center

Adjacent to the VAB is the Launch Control Center (LCC). This four-story structure is a radical departure from the dome-shaped blockhouses at other launch sites.

The electronic "brain" of Launch Complex 39, the LCC was used for checkout and test operations while Apollo 8 was being assembled inside the VAB. The LCC contains display, monitoring, and control equipment used for both checkout and launch operations.

The building has telemeter checkout stations on its second floor, and four firing rooms, one for each high bay of the VAB, on its third floor. Three firing rooms will contain identical sets of control and monitoring equipment, so that launch of a vehicle and checkout of others may take place simultaneously. A ground computer facility is associated with each firing room.

The high speed computer data link is provided between the LCC and the mobile launcher for checkout of the launch vehicle. This link can be connected to the mobile launcher at either the VAB or at the pad.

The three equipped firing rooms have some 450 consoles which contain controls and displays required for the checkout process. The digital data links connecting with the high bay areas of the VAB and the launch pads carry vast amounts of data required during checkout and launch.

There are 15 display systems in each LCC firing room, with each system capable of providing digital information instantaneously.

Sixty television cameras are positioned around the Apollo/Saturn V transmitting pictures on 10 modulated channels. The LCC firing room also contains 112 operational intercommunication channels used by the crews in the checkout and launch countdown.

Mobile Launcher

The mobile launcher is a transportable launch base and umbilical tower for the space vehicle. Three launchers are used at Complex 39.

The launcher base is a two-story steel structure, 25 feet high, 160 feet long, and 135 feet wide. It is positioned on six steel pedestals 22 feet high when in the VAB or at the launch pad. At the launch pad, in addition to the six steel pedestals, four extendible columns also are used to stiffen the mobile launcher against rebound loads, if the engine cuts off.

The umbilical tower, extending 398 feet above the launch platform, is mounted on one end of the launcher base. A hammerhead crane at the top has a hook height of 376 feet above the deck with a traverse radius of 85 feet from the center of the tower.

The 12-million-pound mobile launcher stands 445 feet high when resting on its pedestals. The base, covering about half an acre, is a compartmented structure built of 25-foot steel girders.

The launch vehicle sits over a 45-foot-square opening which allows an outlet for engine exhausts into a trench containing a flame deflector. This opening is lined with a replaceable steel blast shield, independent of the structure, and will be cooled by a water curtain initiated two seconds after liftoff.

There are nine hydraulically operated service arms on the umbilical tower. These swing arms support lines for the vehicle umbilical systems and provide access for personnel to the stages as well as the astronaut crew to the spacecraft.

On Apollo 8 two of the service arms (including the Apollo spacecraft access arm) are retracted early in the count. A third is released at T-30 seconds, and a fourth at about T-15 seconds. The remaining five arms are set to swing back at vehicle first motion after T-0.

The swing arms are equipped with a backup retraction system in case the primary mode fails.

The Apollo access arm (swing arm No. 9), located at the 320 foot level above the launcher base, provides access to the spacecraft cabin for the closeout team and astronaut crews. Astronauts Borman, Lovell and Anders will board the spacecraft starting at about T-2 hours, 40 minutes in the count. The access arm will be moved to a parked position, 12 degrees from the spacecraft, at about T-42 minutes.

This is a distance of about three feet, which permits a rapid reconnection of the arm to the spacecraft in the event of an emergency condition. The arm is fully retracted at the T-5 minute mark in the count.

The Apollo 8 vehicle is secured to the mobile launcher by four combination support and hold-down arms mounted on the launcher deck. The hold-down arms are cast in one piece, about 6 by 9 feet at the base and 10 feet tall, weighing more than 20 tons. Damper struts secure the vehicle near its top.

After the engines ignite, the arms hold Apollo 8 for about six seconds until the engines build up to 95 percent thrust and other monitored systems indicate they are functioning properly. The arms release on receipt of a launch commit signal at the zero mark in the count.

The Transporter

The six-million-pound transporters, the largest tracked vehicles known, move mobile launchers into the VAB and mobile launchers with assembled Apollo Space vehicles to the launch pad. They also are used to transfer the mobile service structure to and from the launch pads. Two transporters are in use at Complex 39.

The Transporter is 131 feet long and 114 feet wide. The vehicle moves on four double-tracked crawlers, each 10 feet high and 40 feet long. Each shoe on the crawler tracks seven feet six inches in length and weighs about a ton. Sixteen traction motors powered by four 1,000-kilowatt generators, which in turn are driven by two 2,750-horsepower diesel engines, provide the motive power for the transporter. Two 750-kw generators, driven by two 1,065-horsepower diesel engines, power the jacking, steering, lighting, ventilating and electronic systems.

Maximum speed of the transporter is about one-mile-per-hour loaded and about two-miles-per-hour unloaded. A 3½ mile trip to the pad with a mobile launcher, made at less than maximum speed, takes approximately seven hours.

The transporter has a leveling system designed to keep the top of the space vehicle vertical within plus-or-minus 10 minutes of arc — about the dimensions of a basketball.

This system also provides leveling operations required to negotiate the five per cent ramp which leads to the launch pad, and keeps the load level when it is raised and lowered on pedestals both at the pad and within the VAB.

The overall height of the transporter is 20 feet from ground level to the top deck on which the mobile launcher is mated for transportation. The deck is flat and about the size of a baseball diamond (90 by 90 feet).

Two operator control cabs, one at each end of the chassis located diagonally opposite each other, provide totally enclosed stations from which all operating and control functions are coordinated.

The transporter moves on a roadway 131 feet wide, divided by a median strip. This is almost as broad as an eight-lane turnpike and is designed to accommodate a combined weight of about 18 million pounds. The roadway is built in three layers with an average depth of seven feet. The roadway base layer is two-and-one-half feet of hydraulic fill compacted to 95 per cent density. The next layer consists of three feet of crushed rock packed to maximum density, followed by a layer of one foot of selected hydraulic fill. The bed is topped and sealed with an asphalt prime coat.

On top of the three layers is a cover of river rock, eight inches deep on the curves and six inches deep on the straightway. This layer reduces the friction during steering and helps distribute the load on the transporter bearings.

Mobile Service Structure

A 402-foot-tall, 9.8-million-pound tower is used to service the Apollo launch vehicle and spacecraft at the pad. The 40-story steel-trussed tower, called a mobile service structure, provides 360-degree platform access to the Saturn vehicle and the Apollo spacecraft.

The service structure has five platforms — two self propelled and three fixed, but movable. Two elevators carry personnel and equipment between work platforms. The platforms can open and close around the 363-foot space vehicle.

After depositing the mobile launcher with its space vehicle on the pad, the transporter returns to a parking area about 7,000 feet from the pad. There it picks up the mobile service structure and moves it to the launch pad. At the pad, the huge tower is lowered and secured to four mount mechanisms.

The top three work platforms are located in fixed positions which serve the Apollo spacecraft. The two lower movable platforms serve the Saturn V.

The mobile service structure remains in position until about T-11 hours when it is removed from its mounts and returned to the parking area.

Water Deluge System

A water deluge system will provide a million gallons of industrial water for cooling and fire prevention during launch of Apollo 8. Once the service arms are retracted at liftoff, a spray system will come on to cool these arms from the heat of the five Saturn F-I engines during liftoff.

On the deck of the mobile launcher are 29 water nozzles. This deck deluge will start immediately after liftoff and will pour across the face of the launcher for 30 seconds at the rate of 50,000 gallons-per-minute. After 30 seconds, the flow will be reduced to 20,000 gallons-per-minute.

Positioned on both sides of the flame trench are a series of nozzles which will begin pouring water at 8,000 gallons-per-minute, 10 seconds before liftoff. This water will be directed over the flame deflector.

Other flush mounted nozzles, positioned around the pad, will wash away any fluid spill as a protection against fire hazards.

Water spray systems also are available along the egress route that the astronauts and closeout crews would follow in case an emergency evacuation was required.

Flame Trench and Deflector

The flame trench is 58 feet wide and approximately six feet above mean sea level at the base. The height of the trench and deflector is approximately 42 feet.

The flame deflector weighs about 1.3 million pounds and is stored outside the flame trench on rails. When it is moved beneath the launcher, it is raised hydraulically into position. The deflector is covered with a four-and-one-half-inch thickness of refractory concrete consisting of a volcanic ash aggregate and a calcium aluminate binder. The heat and blast of the engines are expected to wear about three-quarters of an inch from this refractory surface during the Apollo 8 launch.

Pad Areas

Both Pad A and Pad B of Launch Complex 39 are roughly octagonal in shape and cover about one fourth of a square mile of terrain.

The center of the pad is a hardstand constructed of heavily reinforced concrete. In addition to supporting the weight of the mobile launcher and the Saturn V vehicle, it also must support the 9.8-million-pound mobile service structure and 6-million-pound transporter, all at the same time. The top of the pad stands some 48 feet above sea level.

Saturn V propellants — liquid oxygen, liquid hydrogen, and RP-I — are stored near the pad perimeter.

Stainless steel, vacuum-jacketed pipes carry the liquid oxygen (LOX) and liquid hydrogen from the storage tanks to the pad, up the mobile launcher, and finally into the launch vehicle propellant tanks.

LOX is supplied from a 900,000-gallon storage tank. A centrifugal pump with a discharge pressure of 320 pounds - per square-inch pumps LOX to the vehicle at flow rates as high as 10,000-gallons-per-minute.

Liquid hydrogen, used in the second and third stages, is stored in an 850,000-gallon tank, and is sent through 1,500 feet of 10-inch, vacuum-jacketed invar pipe. A vaporizing heat exchanger pressurizes the storage tank to 60 psi for a 10,000-gallons-per-minute flow rate.

The RP-I fuel, a high grade of kerosene is stored in three tanks — each with a capacity of 86,000 gallons. It is pumped at a rate of 2,000 gallons-per-minute at 175 psig.

The Complex 39 pneumatic system includes a converter compressor facility, a pad high-pressure gas storage battery, a high-pressure storage battery in the VAB, low and high-pressure, cross-country supply lines, high-pressure hydrogen storage and conversion equipment, and pad distribution piping to pneumatic control panels. The various purging systems require 187,000 pounds of liquid nitrogen and 21,000 gallons of helium.

MISSION CONTROL CENTER

The Mission Control Center at the Manned Spacecraft Center, Houston, is the focal point for all Apollo flight control activities. The Center will receive tracking and telemetry data from the Manned Space Flight Network. These data will be processed through the Mission Control Center Real-Time Computer Complex and used to drive displays for the flight controllers and engineers in the Mission Operations Control Room and staff support rooms.

The Manned Space Flight Network tracking and data acquisition stations link the flight controllers at the Center to the spacecraft. For Apollo 8, all stations will be remote sites without flight control teams. All uplink commands and voice communications will originate from Houston, and telemetry data will be sent back to Houston at high speed (2,400 bits per second), on two separate data lines. They can be either real time or playback information.

Signal flow for voice circuits between Houston and the remote sites is via commercial carrier, usually satellite, wherever possible using leased lines which are part of the NASA Communications Network.

Commands are sent from Houston to NASA's Goddard Space Flight Center, Greenbelt Md., lines which link computers at the two points. The Goddard computers provide automatic switching facilities and speed buffering for the command data. Data are transferred from Goddard to remote sites on high speed (2,400 bits per second) lines. Command loads also can be sent by teletype from Houston to the remote sites at 100 words per minute. Again, Goddard computers provide storage and switching functions.

Telemetry data at the remote site are received by the RF receivers, processed by the Pulse Code Modulation ground stations, and transferred to the 642B remote-site telemetry computer for storage. Depending on the format selected by the telemetry controller at Houston, the 642B will output the desired format through a 2010 data transmission unit which provides parallel to serial conversion, and drives a 2,400 bit-per-second modem.

The data modem converts the digital serial data to phase-shifted keyed tones which are fed to the high speed data lines of the Communications Network.

Telemetry summary messages can also be output by the 642B computer, but these messages are sent to Houston on 100 - word - per - minute teletype lines rather than on the high-speed lines.

Tracking data are output from the sites in a low speed (100 words) teletype format and a 240-bit block high speed (2,400 bits) format. Data rates are 1 sample-6 seconds for teletype and 10 samples (frames) per second for high speed data.

All high-speed data, whether tracking or telemetry, which originate at a remote site are sent to Goddard on high-speed lines. Goddard reformats the data when necessary and sends them to Houston in 600-bit blocks at a 40,800 bits-per-second rate. Of the 600-bit block, 480 bits are reserved for data, the other 120 bits for address, sync, intercomputer instructions, and polynominal error encoding.

All wideband 40,800 bits-per-second data originating at Houston are converted to high speed (2,400 bits-per second) data at Goddard before being transferred to the designated remote site.

MANNED SPACE FLIGHT NETWORK

The Manned Space Flight Network (MSFN) will have 14 ground stations, four instrumented ships, and six instrumented aircraft ready for participation in Apollo 8.

The MSFN is designed to keep in close contact with the spacecraft and astronauts at all times, except for the approximate 45 minutes Apollo will be behind the Moon. The network is designed to provide reliable, continuous, and instantaneous communications with the astronauts, launch vehicle, and spacecraft from liftoff to splashdown.

As the spacecraft lifts off from Kennedy Space Center, the tracking stations will be watching it. As the Saturn ascends, voice and data will be instantaneously transmitted to Mission Control Center (MCC) in Houston. Data will be run through computers at MCC for visual display for flight controllers.

Depending on the launch azimuth, a string of 30-foot diameter antennas around the Earth will keep tabs on Apollo 8 and transmit information back to Houston. First, the station at Merritt Island, then it will be Grand Bahama Island, Bermuda, the Vanguard tracking ship, and Canary Island. Later, Carnarvon, Australia, will pick up Apollo 8, followed by Hawaii, the Redstone tracking ship, Guaymas, Mexico, and Corpus Christi, Texas.

For injection into translunar orbit, MCC sends a signal through one of the land stations or one of the three Apollo ships in the Pacific. As the spacecraft heads for the Moon, the engine burn is monitored by the ships and an Apollo/Range Instrumentation Aircraft (A/RIA). The A/RIA provides a relay for the astronauts' voice and data communication with Houston.

As the spacecraft moves away from Earth, first the smaller 30-foot diameter antennas communicate with the spacecraft, then at a spacecraft altitude of 10,000 miles they hand over the tracking function to the larger and more powerful 85-foot antennas. These 85-foot antennas are near Madrid, Spain; Goldstone, Calif.; and Canberra, Australia. The 85-foot antennas are at 120-degree intervals around Earth so at least one antenna has the Moon in view at all times. As the Earth revolves from west to east, one station hands over control to the next station as it moves into view of the spacecraft. In this way, a continuous data and communication flow is maintained.

Data is constantly relayed back through the huge antennas and transmitted via the NASA

Communications Network — a half million miles of land and underseas cables and radio circuits, including those through communications satellites — to MCC. This data is fed into computers for visual display in Mission Control. For example, a display would show on a large map, the exact position of the spacecraft. Or returning data could indicate a drop in power or some other difficulty which would result in a red light going on to alert a Flight Controller to make a decision and take action.

Returning data flowing through the Earth stations give the necessary information for commanding mid-course maneuvers to keep the Apollo in a proper trajectory for orbiting the Moon. On reaching the vicinity of the Moon the data indicate the amount of burn necessary for the service module engine to place the spacecraft in lunar orbit. And so it goes, continuous tracking and acquisition of data between Earth and Apollo are used to fire the spacecraft's engine to return home and place it on the precise trajectory for reentering the Earth's atmosphere.

As the spacecraft comes toward Earth at about 25,000 miles per hour, it must reenter at the proper angle. Calculations based on data coming in at the various tracking stations and ships are fed into the computers at MCC where flight controllers make decisions that will provide the returning spacecraft with the necessary information to make accurate reentry. Appropriate MSFN stations, including tracking ships and aircraft repositioned in the Pacific for this event, are on hand to provide support during reentry. An A/RIA aircraft will relay astronaut voice communications to MCC and antennas on reentry ships will follow the spacecraft.

During the journey to the Moon and back, television will be received from the spacecraft at the various 85-foot antennas around the world: Spain, Goldstone, and Australia, scan converters at Madrid and Goldstone permit immediate transmission via NASCOM to Mission Control where it will be released to TV networks.

NASA Communications Network - Goddard

This network consists of several systems of diversely routed Communications Channels leased on communications satellites, common carrier systems and high frequency radio facilities where necessary to provide the access links.

The system consists or both narrow and wide-band channels, and some TV channels. Included are a variety of telegraph, voice and data systems (digital and analog) with a wide range of digital data rates. Wide-band systems do not extend overseas. Alternate routes or redundancy are provided for added reliability in critical mission operations.

A primary switching center and intermediate switching and control points are established to provide centralized facility and technical control, and switching operations under direct NASA control. The primary switching center is at Goddard, and intermediate switching centers are located at Canberra, Australia; Madrid, Spain; London, England; Honolulu, Hawaii; Guam; and Cape Kennedy, Florida.

For Apollo 8, Cape Kennedy is connected directly to the Mission Control Center, Houston, by the communication network's Apollo Launch Data System, a combination of data gathering and transmission systems designed to handle launch data exclusively.

After launch, all network and tracking data are directed to the Mission Control Center, Houston, through Goddard. A high-speed data line (2,400 bits-per-second) connects Cape Kennedy to Goddard, where the transmission rate is increased to 40,800 bits-per-second from there to Houston. Upon orbital insertion, tracking responsibility is transferred between the various stations as the spacecraft circles the Earth.

Two Intelsat communications satellites will be used for Apollo 8. The Atlantic satellite will service the Ascension Island Unified S-Band (USB) station, the Atlantic Ocean ship and the Canary Island site.

The second Apollo Intelsat communications satellite, over the mid-Pacific, will service the Carnarvon, Australia USB site and the Pacific ocean ships. All these stations will be able to transmit simultaneously through the satellite to Houston via Brewster Flat, Washington, and the Goddard Space Flight Center.

Network Computers

At fraction-of-a-second intervals, the network's digital data processing systems, with NASA's Manned Spacecraft Center as the focal point, "talk" to each other or to the spacecraft in real time. High-speed computers at the remote site (tracking ships included) issue commands or "up" data on such matters as control of cabin pressure, orbital guidance commands, or "go, no-go" indications to perform certain functions.

In the case of information originating from Houston, the computers refer to their pre-programmed information for validity before transmitting the required data to the capsule.

Such "up" information is communicated by ultra high-frequency radio at about 1,200 bits-per-second. Communication between remote ground sites, via high-speed communications links, occurs about the same rate. Houston reads information from these ground sites at 2,400 bits-per-second, as well as from remote sites at 100 words-per-minute.

The computer systems perform many other functions including:
Assuring the quality of the transmission lines by continually exercising data paths.
Verifying accuracy of the messages by repetitive operations.
Constantly updating the flight status.

For "down" data, sensors built into the spacecraft continually sample cabin temperature, pressure, physical information on the astronauts such as heartbeat and respiration, among other items. These data are transmitted to the ground stations at 51.2 kilobits (12,800 binary digits) per second.

At MCC the Computers:
Detect and select changes or deviations, compare with their stored programs, and indicate the problem areas or pertinent data to the flight controllers.
Provide displays to mission personnel.
Assemble output data in proper formats.
Log data on magnetic tape for replay.
Provide storage for "on-call" display for the flight controllers.
Keep time.

Fourteen land stations are outfitted with computer systems to relay telemetry and command information between Houston and Apollo spacecraft: Canberra and Carnarvon, Australia; Guam; Kauai, Hawaii; Goldstone, California; Corpus Christi, Texas; Cape Kennedy, Florida; Grand Bahama Island; Bermuda; Madrid; Grand Canary Island; Antigua; Ascension Island; and Guaymas, Mexico.

Network Configuration for Apollo 8
Unified S-Band (USB) Sites:

NASA 30-Foot Antenna Sites	NASA 85-Foot Antenna Sites
Antigua (ANG)	Canberra (CNB), Australia (Prime)
Ascension Island (ACN)	Goldstone (GDS), California (Prime)
Bermuda (BDA)	Madrid (MAD), Spain (Prime)
Canary Island (CYI)	*Canberra (DSS-42 Apollo Wing) (Backup)
Carnarvon (CRO), Australia	*Goldstone (DSS-II Apollo Wing)
Grand Bahama Island (GBM)	*Madrid (DSS-61 Apollo Wing)
Guam (GWM)	
Guaymas (GYM), Mexico	
Hawaii (HAW) (Backup)	
Merritt Island (MIL), Florida	
Corpus Christie (TEX), Texas (Backup)	

Tananarive (TAN), Malagasy Republic (STADAN station in support role only.)
*Wings have been added to JPL Deep Space Network site operations buildings.
These wings contain additional Unified S-Band equipment as backup to the Prime sites.

APOLLO 8 RECOVERY

The Apollo Ships.
The mission will be supported by four Apollo instrumentation ships operating as integral stations of the Manned Space Flight Network (MSFN) to provide coverage in areas beyond the range of land stations.

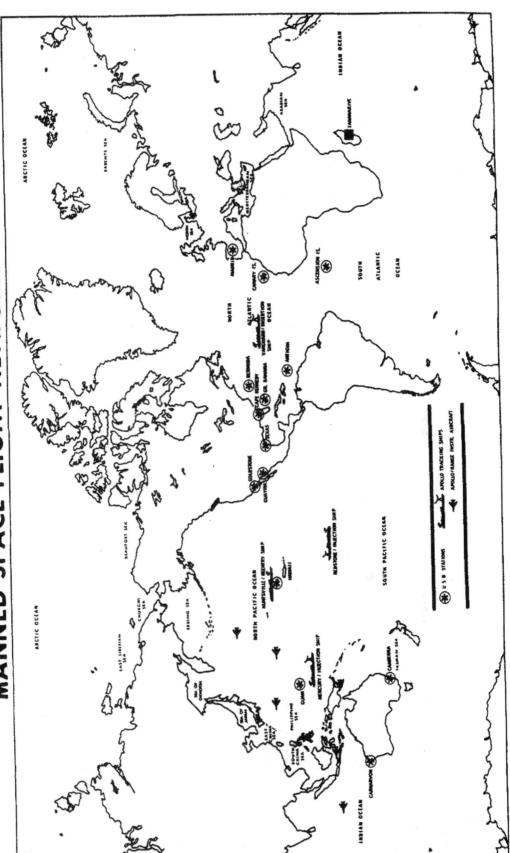

MANNED SPACE FLIGHT NETWORK (APOLLO-8)

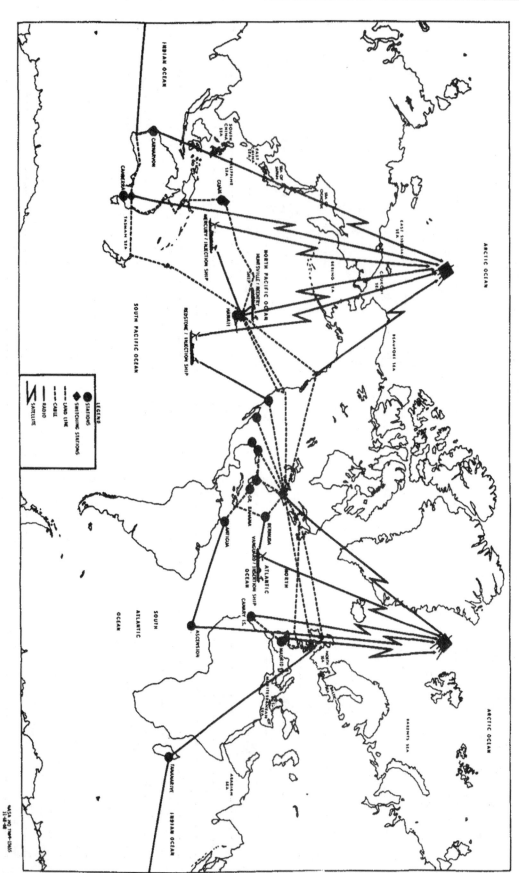

NASCOM–APOLLO 8

LEGEND

STATIONS
SWITCHING STATIONS
CABLE
LAND LINE
RADIO
SATELLITE

NETWORK CONFIGURATION FOR APOLLO 8 MISSION

Facilities	C-band (High Speed)	C-band (Low Speed)	ODOP	Optical	USB	Voice (A/G)	Command	Telemetry	VHF Links	FM Remoting	Mag Tape Recording	Decoms	Displays	CMD Destruct	642B TLM	642B CMD	1218	High Speed Data	Wideband Data	TTY	Voice (SCAMA)	VHF A/G Voice	Video (TV)	SPAN
	Tracking				USB				TLM					CMD	Data Processing			Comm						Other
CIF											X	X						X	X		X		X	
TEL. 4									X	X	X													
CNV	X		X	X										X									X	
PAT*	X	X																						
MLA	X	X																						
MIL	X*	X*			X	X	X	X	X	X	X	X	X		X	X	X	X		X	X	X	X	
GBI	X*	X*					X			X				X										
GBM					X	X	X	X	X	X	X	X			X	X	X	X		X	X	X	X	
GTK	X	X												X										
ANG					X	X	X	X	X	X	X	X			X	X	X	X		X	X	X	X	
ANT	X	X							X		X													
BDA	X	X			X	X	X	X	X	X	X	X	X	X	X	X	X	X		X	X	X	X	
ACN					X	X	X	X	X	X	X	X			X	X	X			X	X	X	X	
ASC		X																						
MAD					X	X	X	X		X	X	X			X	X	X	X		X	X		X	
MADX					X	X	X	X									X			X	X			
CYI		X			X	X	X	X	X	X	X	X			X	X	X	X		X	X	X	X	X
PRE		X																						
TAN		X							X		X									X	X	X		
CRO	X	X			X	X	X	X	X	X	X	X			X	X	X	X		X	X	X	X	X
HSK					X	X	X	X		X	X	X			X	X	X	X		X	X		X	
HSKX					X	X	X	X									X			X	X			
GWM					X	X	X	X	X	X	X	X			X	X	X	X		X	X	X	X	
HAW		X			X	X	X	X	X	X	X	X	X		X	X	X	X		X	X	X	X	
CAL	X	X																		X	X	X		
WHS	X	X																		X	X			
GDS					X	X	X	X		X	X	X			X	X	X	X		X	X		X	
GDSX					X	X	X	X									X			X	X			
GYM					X	X	X	X	X	X	X	X			X	X	X	X		X	X	X	X	
TEX					X	X	X	X	X	X	X	X			X	X	X	X		X	X		X	
HTV		X			X	X		X			X	X								X	X			
RED	X	X			X	X	X	X	X	X	X	X			X	X		X	X	X	X	X	X	
VAN	X	X			X	X	X	X	X	X	X				X	X		X		X	X	X	X	
MER	X	X			X	X	X	X	X	X	X				X	X		X		X	X	X	X	X
ARIA (6)					X	X		X	X		X										X	X		

*Subject to availability.

The ships, Vanguard, Redstone, Mercury, and Huntsville will perform tracking, telemetry, and communication functions for the launch phase, Earth orbit insertion, translunar injection (TLI), and reentry at the end of the mission.

Vanguard will be stationed about 1,000 miles southeast of Bermuda (25°N, 49°W) to bridge the Bermuda-Antigua gap during Earth orbit insertion. Vanguard also functions as part of the Atlantic recovery fleet in the event of a launch phase contingency. Redstone, in the western Pacific, north of Bougainville (2.5°N, 155.5°E); Mercury) 1500 miles further east (7.5°N, 181.5°E); and Huntsville, near Wake Island (21.0°N, 169.0°E), provide a triangle of mobile stations between the MSFN stations at Carnarvon and Hawaii for coverage of the burn interval for translunar injection. In the event the launch date slips from December 21, the ships will all move generally southwestward to cover the changing flight window patterns.

Mercury and Huntsville will be repositioned along the reentry corridor for tracking, telemetry, and communications functions during reentry and landing.

The Apollo ships were developed jointly by NASA and the Department of Defense. The DOD operates the ships in support of Apollo and other NASA and DOD missions on a non-interference basis with Apollo requirements.

The overall management of the Apollo ships is the responsibility of the Commander, Air Force Western Test Range (AFWTR). The Military Sea Transport Service provides the maritime crews and the Federal Electric Corporation of International Telephone and Telegraph, under contract to AFWTR, provides the technical instrumentation crews.

The technical crews operate in accordance with Joint NASA/DOD standards and specifications which are compatible with MSFN operational procedures.

Apollo/Range Instrumentation Aircraft (A/RIA)

The Apollo/Range Instrumentation Aircraft (A/RIA) will support the mission by filling gaps in both land and ship station coverage where important and significant coverage requirements exist.

During Apollo 8, the A/RIA will be used primarily to fill coverage gaps of the land and ship stations in the Pacific during the translunar injection interval (TLI). Prior to and during the TLI burn, the A/RIA record telemetry data from Apollo and provide a real-time voice communication between the astronauts and the flight director at Houston.

Six aircraft will participate in this mission flying from Pacific air bases to positions under the orbital track of the spacecraft and booster.

The A/RIA will fly, initially, out of Hawaii, Guam, and the Philippines, as well as three bases in Australia: Townsville, Darwin, and Perth. The aircraft, like the tracking ships, will also be re-deployed in a southwest direction in the event of launch day slips.

The total A/RIA fleet for Apollo missions consist of eight EC-135-A (Boeing 707) jet aircraft equipped specifically to meet mission needs. Seven-foot parabolic antennas have been installed in the nose section of the aircraft giving them a large, bulbous look.

They are under the overall supervision of the Office of Tracking and Data Acquisition with direct supervision the responsibility of Goddard. The aircraft, as well as flight and instrumentation crews, are provided by the Air Force and they are equipped through joint Air Force NASA contract action.

APOLLO 8 CREW

Crew Training

The crewmen of Apollo 8 have spent more than seven hours of formal crew training for each hour of the lunar - orbit mission's six-day duration. Almost 1,100 hours of training were in the Apollo 8 crew training syllabus over and above the normal preparations for the mission-technical briefings and reviews, pilot meetings and study.

The Apollo 8 crewmen also participated in spacecraft manufacturing checkouts at the North American Rockwell plant in Downey, Calif., and in pre-launch testing at NASA Kennedy Space Center. Taking part in factory and launch area testing has provided the crew with valuable operational knowledge of the complex vehicle.

Highlights of specialized Apollo 8 crew training topics are:
* Detailed series of briefings on spacecraft systems, operation and modifications.

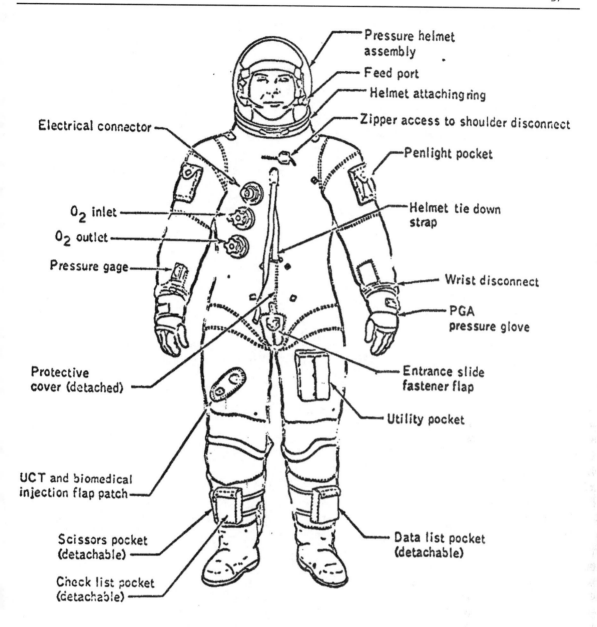

Pressure helmet assembly

Feed port

Helmet attaching ring

Zipper access to shoulder disconnect

Electrical connector

Penlight pocket

Helmet tie down strap

O_2 inlet

O_2 outlet

Pressure gage

Wrist disconnect

PGA pressure glove

Protective cover (detached)

Entrance slide fastener flap

Utility pocket

UCT and biomedical injection flap patch

Scissors pocket (detachable)

Data list pocket (detachable)

Check list pocket (detachable)

Intravehicular configuration of the PGA.

* Saturn launch vehicle briefings on countdown, range safety, flight dynamics, failure modes and abort conditions. The launch vehicle briefings were updated periodically.

* Apollo Guidance and Navigation system briefings at the Massachusetts Institute of Technology Instrumentation Laboratory.

* Briefings and continuous training on mission photographic objectives and use of camera equipment.

* Extensive pilot participation in reviews of all flight procedures for normal as well as emergency situations.

* Stowage reviews and practice in training sessions in the spacecraft, mockups, and Command Module simulators allowed the crewmen to evaluate spacecraft stowage of crew-associated equipment.

* More than 200 hours of training per man in Command Module simulators at MSC and KSC, including closed-loop simulations with flight controllers in the Mission Control Center. Other Apollo simulators at various locations were used extensively for specialized Crew training.

* Entry corridor deceleration profiles at lunar return conditions in the MSC Flight Acceleration

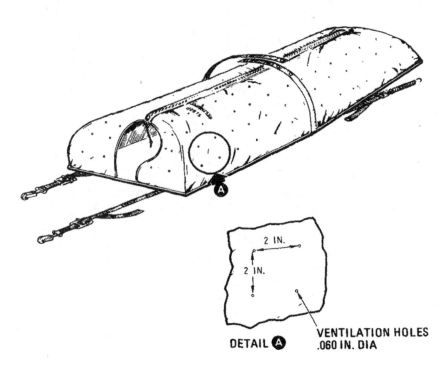

2 IN.

2 IN.

DETAIL Ⓐ

VENTILATION HOLES
.060 IN. DIA

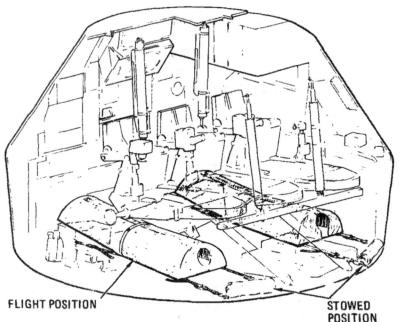

FLIGHT POSITION

STOWED
POSITION

Facility manned centrifuge.

    * Water egress training conducted in indoor tanks as well as in the Gulf of Mexico, included uprighting from the Stable II position (apex down) to the Stable I position (apex up), egress onto rafts and helicopter pickup.

    * Launch pad egress training from mockups and from the actual spacecraft on the launch pad for possible emergencies such as fire, contaminants and power failures.

    * The training covered use of Apollo spacecraft fire suppression equipment in the cockpit.

    * Planetarium reviews at Morehead Planetarium, Chapel Hill, N. C., and at Griffith Planetarium, Los Angeles, Calif., of the celestial sphere with special emphasis on the 37 navigational stars used by the Command Module Computer.

Apollo 8 Spacesuits

Apollo 8 crewmen, until one hour after translunar injection, will wear the intravehicular pressure garment assembly — a multi-layer spacesuit consisting or a helmet, torso and gloves which can be pressurized independently of the spacecraft.

The spacesuit outer layer is Teflon-coated Beta fabric woven of fiberglass strands with a restraint layer, a pressure bladder and an inner high-temperature nylon liner.

Oxygen connection, communications and biomedical data lines attach to fittings on the front of the torso.

A one-piece constant wear garment, similar to "long johns," is worn as an undergarment for the spacesuit and for the in-flight garment is porous-knit cotton with a waist-to-neck zipper for donning. Attach points for the biomedical harness also are provided.

After taking off the spacesuits, the crew will wear Teflon fabric inflight coveralls over the constant wear garment. The two-piece coveralls provide warmth in addition to pockets for personal items. The crew will wear

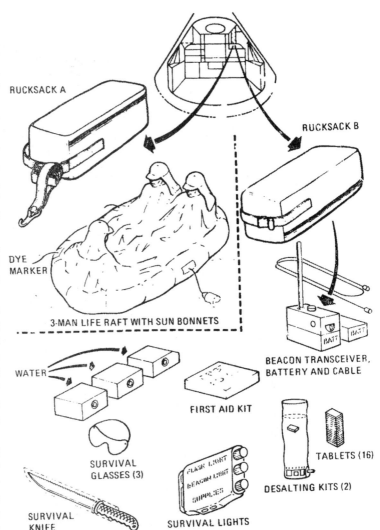

RUCKSACK A

RUCKSACK B

DYE MARKER

3-MAN LIFE RAFT WITH SUN BONNETS

WATER

SURVIVAL GLASSES (3)

SURVIVAL KNIFE

FIRST AID KIT

SURVIVAL LIGHTS

BEACON TRANSCEIVER, BATTERY AND CABLE

TABLETS (16)

DESALTING KITS (2)

the inflight coveralls during entry. The soles of the garment have been fitted with a special metal heel clip which fits in the couch heel restraint. Additionally, fitted fluorel foam pads on couch headrests will provide head restraint during entry. These pads will be stowed until just prior to entry.

The crewmen will wear communications carriers inside the pressure helmet. The communications carriers provide redundancy in that each has two microphones and two earphones.

A lightweight headset is worn with the inflight coveralls.

Apollo 8 Crew Meals

The Apollo 8 crew had a wide range of food items from which to select their daily mission space menu. More than 60 items comprise the selection list of freeze dried bite-size rehydratable foods.

Average daily value of three meals will be 2,500 calories per man.

Unlike Gemini crewmen who prepared their meals with cold water, Apollo crewmen have running water for hot meals and cold drinks.

Water is obtained from three sources — a dispenser for drinking water and two water spigots at the food preparation station, one supplying water at about 155 degrees F., the other at about 55 degrees F. The potable water dispenser emits half-ounce spurts with each squeeze and the food preparation spigots dispense water in one-ounce increments.

Spacecraft potable water is supplied from service module fuel cell by-product water.

The day-by-day, meal-by-meal Apollo 8 menu for each crewman is listed on the following page.

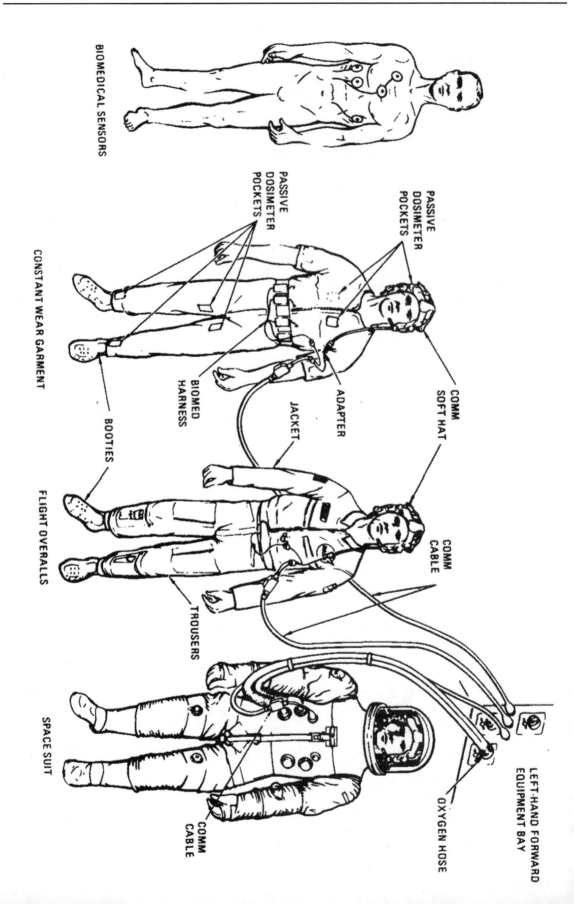

BIOMEDICAL SENSORS

CONSTANT WEAR GARMENT

PASSIVE DOSIMETER POCKETS

PASSIVE DOSIMETER POCKETS

BIOMED HARNESS

BOOTIES

ADAPTER

JACKET

COMM SOFT HAT

FLIGHT OVERALLS

COMM CABLE

TROUSERS

SPACE SUIT

COMM CABLE

OXYGEN HOSE

LEFT-HAND FORWARD EQUIPMENT BAY

Personal Hygiene

Crew personal hygiene equipment aboard Apollo 8 includes body cleanliness items, the waste management system and two medical kits.

Packaged with the food are a toothbrush and a two ounce tube of toothpaste for each crewman. Each man-meal package contains a 3.5 by 4-inch wet-wipe cleansing towel. Additionally, three packages of 12 by 12-inch dry towels are stowed beneath the command module pilot's couch. Each package contains seven towels. Also stowed under the command module pilot's couch are seven tissue dispensers containing 53 3-ply tissues each.

Solid body wastes are collected in Gemini-type plastic defecation bags which contain a germicide to prevent bacteria and gas formation. The bags are sealed after use and stowed in empty food containers for post-flight analysis.

Urine Collection devices are provided for use either while wearing the pressure suit or in the inflight coveralls. The urine is dumped overboard through the spacecraft urine dump valve.

The two medical accessory kits, 6 by 4.5 by 4 inches, are stowed on the spacecraft back wall at the feet of the command module pilot.

The medical kits contain three motion sickness injectors, three pain suppression injectors, one 2-oz bottle first aid ointment, two 1-oz bottle eye drops, three nasal sprays, two compress bandages, 12 adhesive bandages, one oral thermometer and two spare crew biomedical harnesses. Pills in the medical kits are 60 antibiotic, 12 nausea, 12 stimulant, 18 pain killer, 60 decongestant, 24 diarrhea, 72 aspirin and 21 sleeping.

Apollo 8 (Borman, Lovell, Anders)

Day 1* 5 and 9	Day 2, 6, and 10	Day 3, 7, and 11	Day 4, 8, and 12
A. Peaches	Canadian Bacon &	Fruit Cocktail	Canadian Bacon &
Bacon Squares (8)	Applesauce	Bacon Squares (8)	Applesauce
Cinn Tstd Bread Cubes (8)	Sugar Coated Corn	Cinn Tstd Bread Cubes (8)	Toasted Bread
Grapefruit Drink	Flakes	Cocoa	Cubes (8)
	Apricot Cereal Cubes (8)	Orange Drink	Strawberry Cereal
	Grapefruit Drink		Cubes (6)
	Orange Drink		Cocoa
			Orange Drink
B. Corn Chowder	Tuna Salad	Cream of Chicken Soup	Pea Soup
Chicken & Gravy	Chicken & Vegetables	Beef Pot Roast	Chicken & Gravy
Toasted Bread Cubes (6)	Cinn Tstd Bread Cubes (8)	Toasted Bread Cubes (8)	Cheese Sandwiches
Sugar Cookie Cubes (6)	Pineapple Fruitcake (4)	Butterscotch Pudding	(6)
Cocoa	Pineapple-Grapefruit	Grapefruit Drink	Bacon Squares
Orange Drink	Drink		(6)
			Grapefruit Drink
C. Beef & Gravy	Spaghetti & Meat Sauce	Potato Soup	Shrimp Cocktail
Beef Sandwiches (4)	Beef Bites (6)	Chicken Salad	Beef Hash
Cheese-Cracker Cubes (8)	Bacon Squares (6)	Turkey Bites (6)	Cinn Tstd Bread
Chocolate Pudding	Banana Pudding	Graham Cracker Cubes	Cubes (8)
Orange-Grapefruit Drink	Grapefruit Drink	(6)	Date Fruitcake
		Orange Drink	(4)
			Orange-Grapefruit
			Drink
DAYS TOTAL CALORIES			
2485	2537	2522	2441

*Day 1 consists of Meals B and C only; Day 12 consists of Meal A only. Each crew member will be provided with a total of 33 meals.

Sleep-Work Cycles

At least one crew member will be awake at all times. The normal cycle will be 17 hours of work followed by seven hours of rest. Simultaneous rest periods are scheduled for the command module pilot and the lunar module pilot. When possible, all three crewmen will eat together, with one hour allocated for each meal period.

Sleeping positions in the command module are under the left and right couches, with heads toward the crew hatch. Two lightweight Beta fabric sleeping bags are each supported by two longitudinal straps attaching to lithium hydroxide storage boxes at one end and to the spacecraft pressure vessel inner structure at the other end.

Additional transverse restraint straps have been added to the sleeping bags since the Apollo 7 mission to provide greater sleeping comfort and body restraint in zero-g. The sleeping bags have also been perforated for improved ventilation.

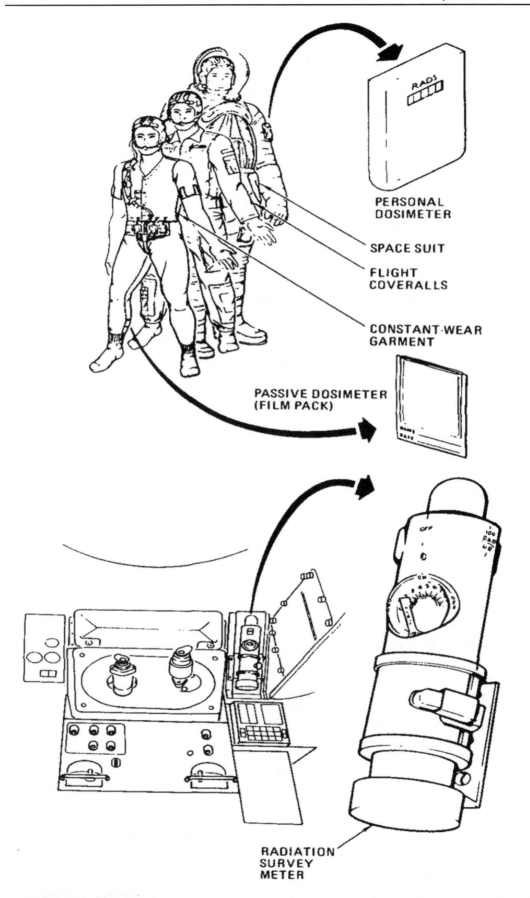

PERSONAL
DOSIMETER

SPACE SUIT

FLIGHT
COVERALLS

CONSTANT·WEAR
GARMENT

PASSIVE DOSIMETER
(FILM PACK)

RADIATION
SURVEY
METER

Survival Gear

The survival kit is stowed in two rucksacks in the right-hand forward equipment bay above the lunar module pilot. Contents of rucksack No. I are: two combination survival lights, one desalter kit, three pair sunglasses, one radio beacon, one spare radio beacon battery and spacecraft connector cable, one machete in sheath, three water containers and two containers of Sun lotion. Rucksack No. 2: one three man life raft with CO2 inflater, one sea anchor, two sea dye markers, three sun bonnets, one mooring lanyard, three man lines and two attach brackets.

The survival kit is designed to provide a 48-hour post landing (water or land) survival capability for three crewmen between 40 degrees North and South Latitudes.

Biomedical Inflight Monitoring

The Apollo 8 crew inflight biomedical telemetry data received by the Manned Space Flight Network will be relayed for instantaneous display at Mission Control Center. Heart rate and breathing rate data will be displayed on the flight surgeon's console. Heart rate and respiration rate average, range and deviation are computed and displayed on the digital TV screens.

In addition, the instantaneous heart rate, real time and delayed EKG and respiration are recorded on strip charts for each man. Biomedical data observed by the flight surgeon and his team in the Life Support Systems Staff Support Room will be correlated with spacecraft and spacesuit environmental data displays. Blood pressure and body temperature are no longer taken as they were in earlier manned flight programs.

The Crew on Launch Day

Following is a timetable of Apollo 8 crew activities on launch day. (All times are shown in hours and minutes before liftoff.)

    T-9:00 - Backup crew alerted
    T-8:30 - Backup crew to LC-39A for spacecraft pre-launch checkouts
    T-5:00 - Flight crew alerted
    T-4:45 - Medical examinations
    T-4:15 - Breakfast
    T-3:45 - Don pressure suits
    T-3:30 - Leave Manned Spacecraft Operations Building for LC-39A via Crew Transfer Van
    T-2:30 - Arrive at LC-39A
    T-2:37 - Enter elevator to spacecraft level
    T-2:40 - Begin spacecraft ingress

Radiation Monitoring

Apollo 8 crew radiation dosages will be closely monitored by onboard dosimeters which either provide crew readouts or telemeter radiation measurements to Manned Space Flight Network stations.

In addition, Solar Particle Alert Network (SPAN) stations will monitor solar flare activity during the mission to provide forecasts of any increase in radiation.

Five types of radiation measuring devices are carried aboard Apollo 8. Each crewman wears standard passive film dosimeters in the thigh, chest and ankle area which provide cumulative post-flight dosage readings. Each man also has a personal radiation dosimeter that can be read for cumulative dosage at any time. They are worn on the right thigh or the pressure garment, and by option on either the shoulder or thigh of the constant wear garment after the pressure suits have been doffed.

Radiation dose rate within the spacecraft cabin is measured by the radiation survey meter, a one-and-a-half pound device mounted in the lower equipment bay near the navigation sextants

A Van Allen belt dosimeter mounted on the spacecraft girth frame near the lunar module pilot's head measures and telemeters onboard radiation skin dose rates and depth dose rates to network stations.

Proton and alpha particle rates and energies exterior to the spacecraft are measured and telemetered by the nuclear particle detection system mounted on the service module forward bulkhead in the area covered by the fairing around the CM-SM mating line. SPAN sites keeping tabs on solar flare activity during Apollo 8 will be NASA-operated stations at Manned Spacecraft Centers Carnarvon Australia, and Canary Islands; and Environmental Sciences Services Administration (ESSA) sites at Boulder, Colo., and Culgoora, Australia.

CREW BIOGRAPHIES
NAME: Frank Borman (Colonel, USAF) Commander
BIRTHPLACE AND DATE: Born March 14, 1928, in Gary, Ind., but grew up in Tucson, Ariz. His parents, Mr. and Mrs. Edwin Borman, now reside in Phoenix, Ariz.
PHYSICAL DESCRIPTION: Blond hair; blue eyes; height: 5 feet 10 inches; weight: 163 pounds.
EDUCATION: Received a Bachelor Of Science degree from the United States Military Academy at West Point in 1950 and a Master Of Science degree in Aeronautical Engineering from the California Institute Of Technology, Pasadena, Calif, in 1957. MARITAL STATUS: Married to the former Susan Bugbee Of Tucson, Ariz.; her mother, Mrs. Ruth Bugbee, resides in Tucson, Ariz.
CHILDREN: Fredrick, October 4, 1951; Edwin, July 20, 1953.   OTHER ACTIVITIES: He enjoys hunting and water skiing.
ORGANIZATIONS: Member of the American Institute Of Aeronautics and Astronautics and the Society of Experimental Test Pilots.
SPECIAL HONORS: Awarded the NASA Exceptional Service Medal, Air Force Astronaut Wings, and Air Force Distinguished Flying Cross; recipient of the 1966 American Astronautical Flight achievement Award and the 1966 Air Force Association David C, Schilling Flight Trophy; co-recipient of the 1966 Harmon International Aviation Trophy; and recipient of the California Institute Of Technology Distinguished Alumni Service Award 1966. EXPERIENCE: Borman, an Air Force Colonel, entered the Air Force after graduation from West Point and received his pilot training at Williams Air Force Base, Arizona.

From 1951 to 1956, he was assigned to various fighter squadrons in the United States and the Philippines.. He became an instructor of thermo-dynamics and fluid mechanics at the Military Academy in 1957 and subsequently attended the USAF Aerospace Research Pilots School from which he graduated in 1960. He remained there as an instructor until 1962. He has accumulated over 5,500 hours flying time, including 4,500 hours in jet aircraft.
CURRENT ASSIGNMENT — Colonel Borman was selected as an astronaut by NASA in September 1962. He has performed a variety of special duties, including an assignment as backup command pilot for the Gemini 4 flight and as a member of the Apollo 204 Review Board.

As command pilot of the history-making Gemini 7 mission, launched on Dec 4, 1965, he participated in establishing a number of space "firsts" — among which are the longest manned space flight (330 hours and 35 minutes) and the first rendezvous of two manned maneuverable spacecraft as Gemini 7 was joined in orbit by Gemini 6.

NAME: James A. Lovell, Jr. (Captain, USN) Command Module Pilot
BIRTHPLACE AND DATE: Born March 25, 1928, in Cleveland, Ohio. His mother, Mrs. Blanche Lowell, resides at Edgewater Beach, Fla.
PHYSICAL DESCRIPTION: Blond hair; blue eyes; height: 5 feet 11 inches; weight: 170 pounds.
EDUCATION: Graduated from Juneau High School, Milwaukee, Wisc.; attended the University of Wisconsin for 2 years, then received a Bachelor of Science degree from the United States Naval Academy in 1952. MARITAL STATUS: Married to the former Marilyn Gerlach of Milwaukee, Wisc. Her parents, Mr. and Mrs. Carl Gerlach, are residents of Milwaukee.
CHILDREN: Barbara L., October 13, 1953 James A., February 15, 1955; Susan K., July 14, 1950; Jeffrey C., January 14, 1966. OTHER ACTIVITIES: His hobbies are golf, swimming, handball, and tennis.
ORGANIZATIONS: Member of the Society of Experimental Test Pilots and the Explorers Club.
SPECIAL HONORS: Awarded two NASA Exceptional Service Medals, the Navy Astronaut Wings, two Navy Distinguished Flying Crosses, and the 1967 FAI Gold Space Medal (Athens, Greece); and co-recipient of the 1966 American Astronautical Society Flight Achievement Award and the Harmon International Aviation Trophy in 1966 and 1967. EXPERIENCE: Lovell, a Navy Captain, received flight training following graduation from Annapolis.

He has had numerous naval aviator assignments including a 4-year tour as a test pilot at the Naval Air Test Center, Patuxent River, Md.. While there he served as program manager for the F4H weapon system evaluation. A graduate of the Aviation Safety School of the University of Southern California, he also served as a flight instructor and safety officer with Fighter Squadron 101 at the Naval Air Station, Oceana, Va..

Of the 4,000 hours flying time he has accumulated, more than 3,000 hours are in Jet aircraft.
CURRENT ASSIGNMENT: Captain Lovell was selected as an astronaut by NASA in September 1962. He has since served as backup pilot for the Gemini 4 flight and backup command pilot for the Gemini 9 flight.

On Dec. 4, 1965, he and command pilot Frank Borman were launched into space on the history-making Gemini 7 mission. The flight lasted 330 hours and 35 minutes, during which the following space "firsts" were accomplished: longest manned space flight; first rendezvous of two manned maneuverable spacecraft, as Gemini 7 was joined in orbit by Gemini 6; and longest multi-manned space flight. It was also on this flight that numerous technical and medical experiments were completed successfully. The Gemini 12 mission, with Lovell and pilot Edwin Aldrin, began on Nov. 11, 1966. This 4-day 59-revolution flight brought the Gemini Program to a successful close. Major accomplishments of the 94-hour 35-minute flight included a third-revolution rendezvous with the previously launched Agena (using for the first time backup onboard computations due to a radar failure); a tethered station-keeping exercise; retrieval of a micro-meteorite experiment package from the spacecraft exterior; an evaluation of the use of body restraints specially designed for completing work tasks outside of the spacecraft; and completion of numerous photographic experiments, the highlights of which are the first pictures taken from space of an eclipse of the Sun.

Gemini 12 ended when retrofire occurred at the beginning of the 60th revolution, followed by the second consecutive fully automatic controlled reentry of a spacecraft, and a landing in the Atlantic within 2+ miles of the prime recovery ship USS WASP.

As a result of his participation in this flight, Lovell holds the space endurance record, with 425 hours and 10 minutes, for total time spent in space. Aldrin established a new EVA record by completing 5½ hours outside the spacecraft during two standup EVA's and one umbilical EVA.

SPECIAL ASSIGNMENT: In addition to his regular duties as a member of the astronaut group, Captain Lovell was selected in June 1967 to serve as special Consultant to the President's Council on Physical Fitness.

NAME: William A. Anders (Major, USAF), Lunar Module Pilot
BIRTHPLACE AND DATE: Born October 17, 1933, in Hong Kong; his parents, Commander (USN retired) and Mrs. Arthur F. Anders, now reside in La Mesa, Calif.
PHYSICAL DESCRIPTION: Brown hair; blue eyes; height: 5 feet 8 inches; weight: 145 pounds.
EDUCATION: Received a Bachelor of Science degree from the United States Naval Academy in 1955 and a Master of Science degree in Nuclear Engineering from the Air Force Institute of Technology at Wright-Patterson Air Force Base, Ohio, in 1962.
MARITAL STATUS: Married to the former Valerie E. Hoard of Lemon Grove, Calif., daughter of Mr. and Mrs. Henry G. Hoard, of Oceanside, Calif.
CHILDREN: Alan, February 1957; Glen, July 1958; Gayle, December 1960; Gregory, December 1962; Erie, July 1964.
OTHER ACTIVITIES: His hobbies are fishing, flying, camping, and water skiing; and he also enjoys soccer.
ORGANIZATIONS: Member of the American Nuclear Society and Tau Beta pi.
SPECIAL HONORS: Awarded the Air Force Commendation Medal.
EXPERIENCE: Anders, an Air Force Major, was commissioned in the Air Force upon graduation from the Naval Academy. After Air Force flight training, he served as a fighter pilot in all-weather interceptor squadrons of the Air Defense Command.

After his graduate training, he served as a nuclear engineer and instructor pilot at the Air Force Weapons Laboratory Kirtland Air Force Base, N.M., where he was responsible for technical management of radiation nuclear power reactor shielding and radiation effects programs.

He has logged more than 3,000 hours flying time.
CURRENT ASSIGNMENT: Major Anders was one of the third group of astronauts selected by NASA in October 1963. He has since served as backup pilot for the Gemini 11 mission.

LUNAR DESCRIPTION
Terrain — Mountainous and crater-pitted, the former rising thousands of feet and the latter ranging from a few inches to 180 miles in diameter. The craters are thought to be formed by the impact of meteorites. The surface is covered with a layer of fine-grained material resembling silt or sand, as well as small rocks.

Environment — No air, no wind, and no moisture. The temperature ranges from 250 degrees in the two-week lunar day to 280 degrees below zero in the two-week lunar night. Gravity is one-sixth that of Earth. Micro-meteoroids pelt the Moon (there is no atmosphere to burn them up). Radiation might present a problem during periods of unusual solar activity.

Dark Side — The dark or hidden side of the Moon no longer is a complete mystery. It was first photographed by a Russian craft and since then has been photographed many times, particularly by NASA's Lunar Orbiter spacecraft.

Origin — There is still no agreement among scientists on the origin of the Moon. The three theories: (1) the Moon once was part of Earth and split off into its own orbit, (2) it evolved as a separate body at the same time as Earth, and (3) it formed elsewhere in space and wandered until it was captured by Earth's gravitational field.

Earth to Moon Distances

Date		Surface to Surface
Dec. 21	5 p.m. EST	220,074 statute
Dec. 22	6 p.m. EST	223,337 statute
Dec. 23	7 p.m. EST	227,182 statute
Dec. 24	7:30 p.m. EST	231,238 statute
Dec. 25	8 p.m. EST	235,186 statute
Dec. 26	9 p.m. EST	238,751 statute
Dec. 27	10 p.m. EST	241,779 statute

Physical Facts

Diameter	2,160 miles (about 1 that of Earth)
Circumference	6,790 miles (about ¼ that of Earth)
Distance from Earth	238,857 miles (mean; 221,463 minimum to 252,710 maximum)
Surface temperature	250 (Sun at zenith)-280 (night)
Surface gravity	1/6 that of Earth
Mass	1/100th that of Earth
Volume	1/50th that of Earth
Lunar day and night	14 Earth days each
Mean velocity in orbit	2,287 miles per hour
Escape velocity	1.48 miles per second

Month (period of rotation 27 days, 7 hours, 43 minutes around Earth)

APOLLO PROGRAM MANAGEMENT/CONTRACTORS

Direction of the Apollo Program, the United States' effort to land men on the Moon and return them safely to Earth before 1970, is the responsibility of the Office of Manned Space Flight (OMSF), National Aeronautics and Space Administration, Washington, D.C.

NASA Manned Spacecraft Center (MSC), Houston, is responsible for development of the Apollo spacecraft, flight crew training and flight control.

NASA Marshall Space Flight Center (MSFC), Huntsville, Ala., is responsible for development of the Saturn launch vehicles.

NASA John F. Kennedy Space Center (KSC), Fla., is responsible for Apollo/Saturn launch operations.

NASA Goddard Space Flight Center (GSFC), Greenbelt, Md., manages the Manned Space Flight Network under the direction of the NASA Office of Tracking and Data Acquisition (OTDA).

Apollo/Saturn Officials

Dr. George E. Mueller	Associate Administrator for Manned Space Flight, NASA Headquarters
Maj. Gen. Samuel C. Phillips	Director, Apollo Program Office, OMSF, NASA Headquarters
George H. Hage	Deputy Director, Apollo Program Office, OMSF, NASA Headquarters
William C. Schneider	Apollo Mission Director, OMSF, NASA Headquarters
Chester M. Lee	Assistant Mission Director, OMSF, NASA Headquarters
Col. Thomas H. McMullen	Assistant Mission Director, OMSF, NASA Headquarters
Dr. Robert R. Gilruth	Director, Manned Spacecraft Center, Houston
George M. Low	Manager, Apollo Spacecraft Program, MSC
Kenneth S. Kleinknecht	Manager, Command and Service Modules, Apollo Spacecraft Program Office, MSC
Donald K. Slayton	Director, Flight Crew Operations, MSC
Christopher C. Kraft, Jr.	Director Flight Operations, MSC
Clifford E. Charlesworth	Apollo 8 Flight Directors,
Glynn S. Lunney	Flight Operations, MSC
M.L. Windler	

Dr. Wernher von Braun	Director, Marshall Space Flight Center, Huntsville, Ala.
Brig. Gen. Edmund F. O'Connor	Director, Industrial Operations, MSFC
Lee B. James	Manager, Saturn V Program Office, MSFC
William D. Brown	Manager, Engine Program Office, MSFC
Dr. Kurt H. Debus	Director, John F. Kennedy Space Center, Fla.
Miles Ross	Deputy Director, Center Operations, KSC
Rocco A. Petrone	Director, Launch Operations, KSC
Walter J. Kapryan	Deputy Director, Launch Operations, KSC
Dr. Hans F. Gruene	Director, Launch Vehicle Operations, KSC
Rear Adm. Roderick O. Middleton	Manager, Apollo Program Office, KSC
John J. Williams	Director, Spacecraft Operations, KSC
Paul C. Donnelly	Launch Operations Manager, KSC
Gerald M. Truszynski	Associate Administrator, Tracking and Data Acquisition, NASA Headquarters
H. R. Brockett	Deputy Associate Administrator, OTDA, NASA Headquarters
Norman Pozinsky	Director, Network Support Implementation Division, OTDA
Dr. John F. Clark	Director, Goddard Space Flight, Greenbelt, Md.
Ozro M. Covington	Assistant Director for Manned Space Flight Tracking, GSFC
Henry F. Thompson	Deputy Assistant Director for Manned Space Flight Support, GSFC
B. William Wood	Chief, Manned Flight Operations Division, GSFC
Tecwyn Roberts	Chief, Manned Flight Engineering Division, GSFC
L. R. Stelter	Chief, NASA Communications Division, GSFC
Maj. Gen. Vincent G. Huston	USAF, DOD Manager of Manned Space Flight Support Operations
Maj. Gen. David M. Jones	USAF, Deputy DOD Manager of Manned Space Flight Support Operations, Commander USAF Eastern Test Range
Rear Adm. F. E. Bakutis	USN, Commander Combined Task Force 130 Pacific Recovery Area (Primary)
Rear Adm. P. S. McManus	USN, Commander Combined Task Force 140 Atlantic Recovery Area
Col. Royce G. Olson	USAF, Director, DOD Manned Space Flight office
Brig. Gen. Allison C. Brooks	USAF, Commander Aerospace Rescue and Recovery Service

Major Apollo/Saturn V Contractors
Contractor Item

The Boeing Co. New Orleans	First Stages (SIC) of Saturn V Flight Vehicles, Saturn V Systems Engineering and Integration Ground Support Equipment
North American Rockwell Corp. Space Division Seal Beach, Calif.	Development and Production of Saturn V Second Stage (S-II)
McDonnell Douglas Astronautics Huntington Beach, Calif.	Development and Production Co. (S-IVB) of Saturn V Third Stage
International Business Machines Federal Systems Division Huntsville, Ala.	Instrument Unit (Prime Contractor)
Bendix Corp., Teterboro, N.J.	Guidance Components for Navigation and Control Div. Instrument Unit (Including ST-124M Stabilized Platform)
Trans World Airlines, Inc.	Installation Support, KSC
Federal Electric Corp.	Communications and Instrumentation Support, KSC
Bendix Field Engineering Corp. Support, KSC	Launch Operations/Complex
Catalytic-Dow	Facilities Engineering and Modifications, RSC
ILC Industries Dover, Del.	Space Suits
Radio Corporation of America Van Nuys, Calif.	110A Computer - Saturn Checkout
Sanders Associates Nashua, New Hampshire	Operational Display Systems Saturn
Brown Engineering Huntsville, Alabama	Discrete Controls

Ingalls Iron Works Birmingham, Alabama	Mobile Launchers (structural work)
Bellcomm Washington, D.C.	Apollo Systems Engineering
The Boeing Co. Washington, D.C.	Technical Integration and Evaluation
General Electric-Apollo Support Department, Daytona Beach, Fla.	Apollo Checkout and Reliability
North American Rockwell Corp. Space Division, Downey, Calif.	Spacecraft Command and Service Modules
Grumman Aircraft Engineering Corp., Bethpage, N.Y.	Lunar Module
Massachusetts Institute of Technology, Cambridge, Mass.	Guidance & Navigation (Technical Management)
General Motors Corp., AC Electronics Division, Milwaukee	Guidance & Navigation (Manufacturing)
TRW Systems Inc. Redondo Beach, Calif.	Trajectory Analysis
Avco Corp., Space Systems Division, Lowell, Mass.	Heat Shield Ablative Material
North American Rockwell Corp. Rocketdyne Division Canoga Park, Calif.	J-2 Engines, F-I Engines
Smith/Ernst (Joint venture) Tampa, Florida Washington, D.C.	Electrical Mechanical Portion of MLs
Power Shovel, Inc. Marion, Ohio	Crawler-Transporter
Hayes International Birmingham, Alabama	Swing Arm

APOLLO 8 GLOSSARY

Ablating Materials —	Special heat-dissipating materials on the surface of a spacecraft that can be sacrificed (carried away, vaporized) during re-entry
Abort —	The cutting short of an aerospace mission before it has accomplished its objective.
Accelerometer —	An instrument to sense accelerative forces and convert them into corresponding electrical quantities usually for controlling, measuring, indicating or recording purposes.
Adapter Skirt —	A flange or extension of a stage or section that provides a ready means of fitting another stage or section to it.
Antipode —	Point on surface of planet exactly 180 degrees opposite from reciprocal point on a line projected through center of body. In Apollo 8 usage, antipode refers to a line from the center of the Moon through the center of the Earth and projected to the Earth surface on the opposite side. The antipode crosses the mid-Pacific recovery line along the 165th meridian of longitude once each 24 hours.
Apocynthion —	Point at which object in lunar orbit is farthest from lunar surface — object having been launched from body other than Moon. (Cynthia, Roman goddess of Moon).
Apogee —	The point at which a moon or artificial satellite in its orbit is farthest from Earth.
Apolune —	Point at which object launched from the Moon into lunar orbit is farthest from lunar surface. e.g. Ascent stage of lunar module after staging into lunar orbit following lunar landing.
Attitude —	The position of an aerospace vehicle as determined by the inclination of its axes to some frame of reference; for Apollo, an inertial, space-fixed reference is used.
Burnout —	The point when combustion ceases in a rocket engine.
Canard —	A short, stubby wing-like element affixed to an aircraft or spacecraft to provide better stability.

Celestial Guidance —	The guidance of a vehicle by reference to celestial bodies.
Celestial Mechanics —	The science that deals primarily with the effect of force as an agent in determining the orbital paths of celestial bodies.
Cislunar —	Adjective referring to space between Earth and the Moon, or between Earth and moon's orbit.
Closed Loop —	Automatic control units linked together with a process to form an endless chain.
Control System —	A system that serves to maintain attitude stability during forward flight and to correct deflections.
Deboost —	A retrograde maneuver which lowers either perigee or apogee of an orbiting spacecraft. Not to be confused with *deorbit*.
Delta V —	Velocity change.
Digital Computer —	A computer in which quantities are represented numerically and which can be used to solve complex problems.
Down-Link —	The part of a communication system that receives, processes and displays data from a spacecraft.
Entry Corridor —	The final flight path of the spacecraft before and during Earth re-entry.
Escape Velocity —	The speed a body must attain to overcome a gravitational field, such as that of Earth; the velocity of escape at the Earth's surface is 36,700 feet-per second.
Explosive Bolts —	Bolts surrounded with an explosive charge which can be activated by an electrical impulse.
Fairing —	A piece, part or structure having a smooth, streamlined outline, used to cover a nonstreamlined object or to smooth a junction.
Fuel Cell —	An electrochemical generator in which the chemical energy from the reaction of oxygen and a fuel is converted directly into electricity.
G or G Force —	Force exerted upon an object by gravity or by reaction to acceleration or deceleration, as in a change of direction: one G is the measure of the gravitational pull required to move a body at the rate of about 32.16 feet-per-second.
Gimbaled Motor —	A rocket motor mounted on gimbal; i.e., on a contrivance having two mutually perpendicular axes of rotation, so as to obtain pitching and yawing correction moments.
Guidance System —	A system which measures and evaluates flight information, correlates this with target data, converts the result into the conditions necessary to achieve the desired flight path, and communicates this data in the form of commands to the flight control system.
Heliocentric —	Sun-centered orbit or other activity which has the Sun as its center.
Inertial Guidance —	A sophisticated automatic navigation system using gyroscopic devices, etc., for high-speed vehicles. It absorbs and interprets such data as speed, position, etc., and automatically adjusts the vehicle to a predetermined flight path. Essentially, it knows where it's going and where it is by knowing where it came from and how it got there. It does not give out any signal so it cannot be detected by radar or jammed.
Injection —	The process of injecting a spacecraft into a calculated orbit.
Multiplexing —	The simultaneous transmission of two or more signals within a single channel. The three basic methods of multiplexing involve the separation of signals by time division, frequency division and phase division.
Optical Navigation —	Navigation by sight, as opposed to mathematical methods.
Oxidizer —	In a rocket propellant, a substance such as liquid oxygen or nitric acid that yields oxygen for burning the fuel.
Penumbra —	Semi-dark portion of a shadow in which light is partly cut off. e.g., Surface of Moon or Earth away from Sun. (See <u>umbra</u>.)
Pericynthion —	Point nearest moon of object in lunar orbit — object having been launched from body other than moon.
Perigee —	The point at which a moon or an artificial satellite in its orbit is closest to the Earth.
Perilune —	The point at which a satellite (e.g., a spacecraft) in its orbit is closest to the Moon: differs from pericynthion in that the orbit is Moon-originated.
Pitch —	The movement of a space vehicle about an axis (Y.) that is perpendicular to its longitudinal axis.
Re-entry —	The return of a spacecraft that re-enters the atmosphere after flight above it.
Retrorocket —	A rocket that gives thrust in a direction opposite to the direction of the object's motion.
Roll —	The movements of a space vehicle about its longitudinal (X) axis.
S Band —	A radio-frequency band of 1550 to 5200 megacycles per second.
Selenographic —	Adjective relating to physical geography of Moon. Specifically, positions on lunar surface as measured in latitude from lunar equator and in longitude from a reference lunar meridian.
Selenocentric —	Adjective referring to orbit having Moon as center. (Selene, Gr. moon)
Sidereal —	Adjective relating to measurement of time, position or angle in relation to the celestial sphere and the vernal equinox.
Telemetering —	A system for taking measurements within an aerospace vehicle in flight and transmitting them by radio to a ground station.
Terminator —	Separation line between lighted and dark portions of celestial body which is not self luminous.
Ullage —	The volume in a closed tank or container above the surface of a stored liquid. Also the ratio of this volume to the total volume of the tank.
Umbra —	Darkest part of a shadow in which light is completely absent. e.g., Surface of Moon or Earth away from Sun.
Up-Link Data —	Telemetry information from the ground.
Yaw —	Displacement of a space vehicle from its vertical (Z) axis.

Apollo 8 Acronyms
        (Note: This list makes no attempt to include all Apollo program acronyms. Listed are several acronyms that are encountered for the first time in the Apollo 8 mission.)

AK	Apogee kick
COI	Contingency orbit insertion
EOI	Earth orbit insertion
HGA	High-gain antenna
IRIG	Inertial reference Integrating Gyro
LOI	Lunar orbit insertion
LPO	Lunar parking orbit
LTAB	Lunar (module) test article B
MCC	Midcourse correction, Mission Control Center
MSI	Moon sphere of influence
REFSMMAT	Reference to stable member matrix
TEI	Transearth injection
TEMCC	Transearth midcourse correction
TLI	Translunar injection
TLMCC	Translunar midcourse correction

Conversion Factors

	Multiply	By	To Obtain
Distance:	feet	0.3048	meters
	meters	3.281	feet
	kilometers	3281	feet
	statute miles	1.609	kilometers
	nautical miles	1.852	kilometers
	nautical miles	1.1508	statute miles
	statute miles	0.86898	nautical miles
Velocity:	feet/sec	0.3048	meters/sec
	meters/sec	3.281	feet/sec
	feet/sec	0.6818	statute miles/hr
	statute miles/hr	1.609	km/hr
	km/hr	0.6214	statute miles/hr
Liquid measure, weight:			
	gallons	3.785	liters
	liters	0.2642	gallons
	pounds	0.4536	kilograms
	kilograms	2.205	pounds
	pounds	14.0	stones

Electrical Power Conversion
Voltage X Current in amps = power in watts
Watts + voltage = amps
Watts + amps = voltage

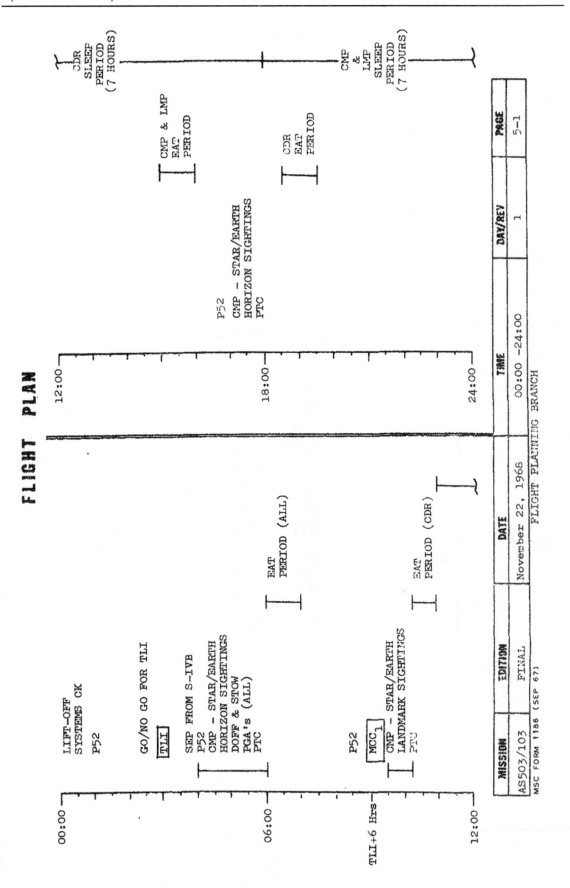

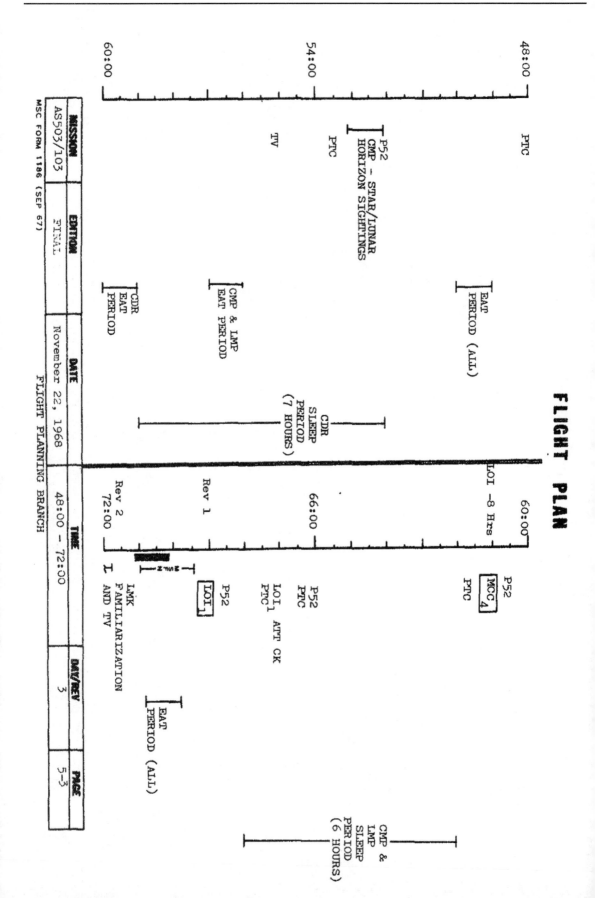

FLIGHT PLAN

MISSION	EDITION	DATE	TIME	DAY/REV	PAGE
AS503/103	FINAL	November 22, 1968	48:00 — 72:00	3	5-3

MSC FORM 1186 (SEP 67)   FLIGHT PLANNING BRANCH

FLIGHT PLAN

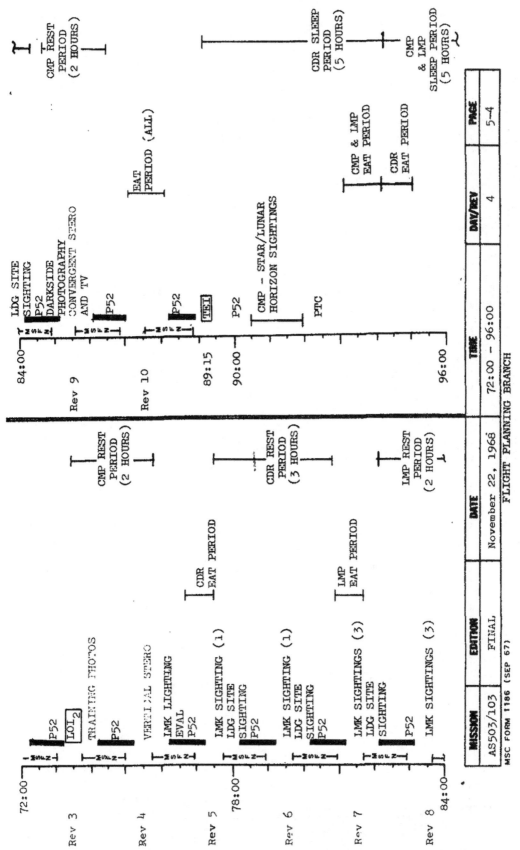

MISSION	EDITION	DATE	DAY/REV	PAGE
AS503/103	FINAL	November 22, 1968	4	5-4

		TIME
		72:00 - 96:00

FLIGHT PLANNING BRANCH

MSC FORM 1186 (SEP 67)

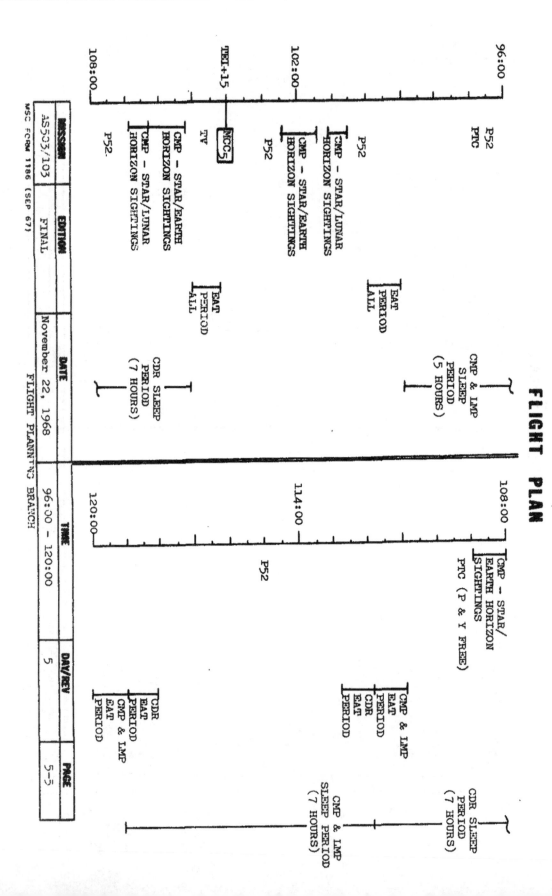

FLIGHT PLAN

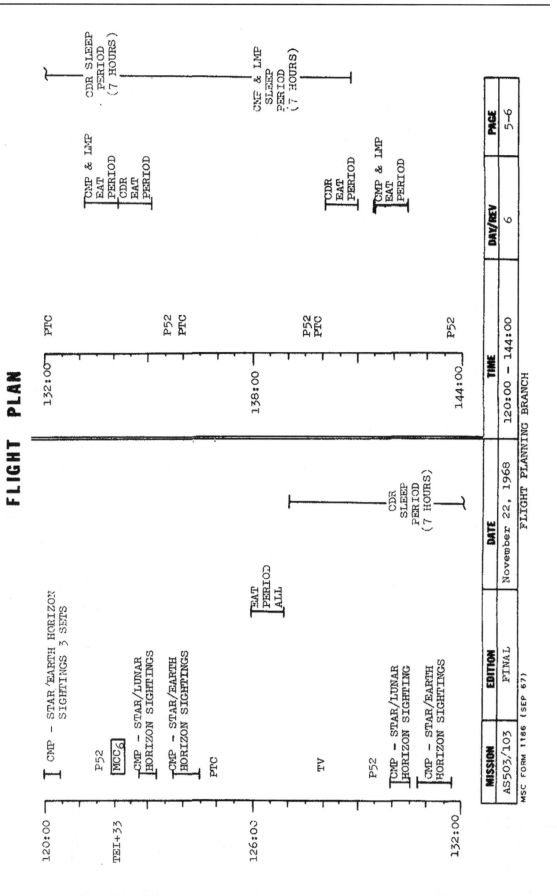

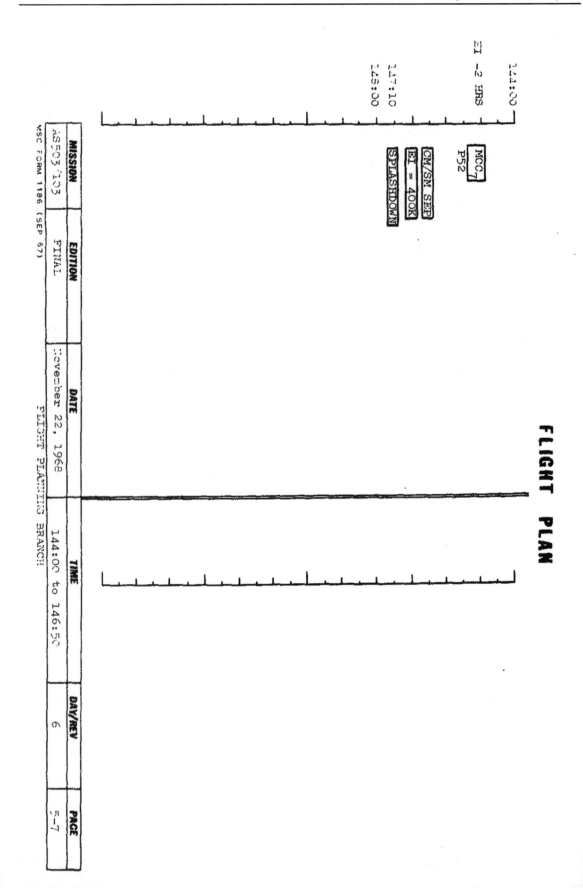

FLIGHT PLAN

MCC₇
P52

CM/SM SEP
EI = 400K
SPLASHDOWN

EI -2 HRS

MISSION	EDITION	DATE	TIME	DAY/REV	PAGE
AS503/103	FINAL	November 22, 1968	144:00 to 146:50	6	5-7

FLIGHT PLANNING BRANCH

MSC FORM 1186 (SEP 67)

Pre - Launch
Mission Operation Report

JAMES A. LOVELL, JR.   WILLIAM A. ANDERS   FRANK BORMAN

MEMORANDUM

To: A/Acting Administrator

From: MA/Apollo Program Director

Subject: Apollo 8 Mission (AS-503)

No earlier than 21 December 1968, we plan to launch the next Apollo/Saturn V mission, Apollo 8. This will be the first manned Saturn V flight, the second flight of a manned Apollo spacecraft and the first manned Apollo flight to the lunar vicinity.

The purpose of this mission is to demonstrate: crew/space vehicle/mission support facilities performance during a manned Saturn V mission with CSM, and performance of nominal and selected backup lunar orbital mission activities including translunar injection, CSM navigation, communications and midcourse corrections, and CSM consumables assessment and passive thermal control. Mission duration is planned for approximately six days and three hours.

The launch will be the third Saturn V from Launch Complex 39 at Kennedy Space Center. The launch window opens at 7:51 EST 21 December and closes for December at 18:20 EST 27 December. Daily windows during this period vary in duration from 4½ hours to approximately 1½ hours.

The nominal mission will comprise: ascent to parking orbit; translunar injection by the S-IVB; CSM separation from S-IVB; translunar coast with required midcourse corrections; lunar orbit insertion and circularization; up to 10 lunar orbits; transearth injection, transearth coast and required midcourse corrections; reentry and splashdown. Recovery will be in the Pacific recovery area with exact location dependent on launch conditions.

APPROVAL:

Sam C. Phillips   George E. Mueller
Lt. General, USAF   Associate Administrator for
Apollo Program Director Manned Space Flight

APOLLO 8 MISSION OPERATION REPORT

The Apollo 8 Mission Operation Report is published in two volumes - I, The Mission Operation Report (MOR) and II, The Mission Operation Report Supplement.

This format was designed to provide a mission oriented document in the MOR with only a very brief description of the space vehicle and support facilities. The MOR Supplement is a reference document with a more comprehensive description of the space vehicle and associated systems,

GENERAL

The goal of the Apollo Program is to enhance the manned space flight capability of the United States by developing, through logical and orderly evolution, the ability to land men an the moon and return them safely to earth.

To accomplish the goal of lunar landing and return in this decade, the Apollo Program has focused on the development of a highly reliable launch vehicle and spacecraft system. This has been done through a logical sequence of Apollo missions designed to qualify the flight hardware, ground support systems and operational personnel in the most effective manner.

The Apollo 8 mission is the second manned flight of the Apollo spacecraft and the first manned Saturn V mission. The mission is designed to test the space vehicle, mission support facilities, and crew on a lunar orbital mission.

PROGRAM DEVELOPMENT

The first Saturn vehicle was successfully flown on 27 October 1961 to initiate operations of the Saturn I Program. A total of 10 Saturn I vehicles (SA-I to SA-10) were successfully flight tested to provide information on the integration of launch vehicle and spacecraft and to provide operational experience with large multi-engined booster stages (S-I, S-IV).

The next generation of vehicles, developed under the Saturn IB Program, featured an uprated first stage (S-IB) and a more powerful new second stage (S-IVB). The first Saturn IB was launched on 26 February 1966. The first three Saturn IB missions (AS-201, AS-203, and AS-202) successfully tested the performance of the launch vehicle and spacecraft combination, separation of the stages, behavior of liquid hydrogen in a weightless environment, performance of the Command Module heat shield at low earth orbital entry conditions, and recovery operations.

The planned fourth Saturn IB mission (AS 204) scheduled for early 1967 was intended to be the first manned Apollo flight. This mission was not flown because of a spacecraft fire, during a manned pre-launch test, that took the lives of the prime flight crew and severely damaged the spacecraft. The SA-204 launch vehicle was later assigned to the Apollo 5 mission.

The Apollo 4 mission was successfully executed on 9 November 1967. This mission initiated the use of the Saturn V launch vehicle (SA-501) and required an orbital restart of the S-IVB third stage. The spacecraft for this mission consisted of an unmanned Command and Service Module (CSM) and a Lunar Module Test Article (LTA). The CSM Service Propulsion System (SPS) was exercised, including restart, and the Command Module Block II heat shield was subjected to the combination of high heat load, high heat rate, and aerodynamic loads representative of lunar return entry.

The Apollo 5 mission was successfully launched and completed on 22 January 1968. This was the fourth mission utilizing Saturn IB vehicles (AS-204). This flight provided for unmanned orbital testing of the Lunar Module (LM-1). The LM structure, staging, proper operation of the Lunar Module Ascent Propulsion System (APS) and Descent Propulsion System (DPS), including restart, was verified. Satisfactory performance of the S-IVB/Instrument Unit (IU) in orbit was also demonstrated.

The Apollo 6 mission (second unmanned Saturn V) was successfully launched on 4 April 1968. Some flight anomalies encountered included oscillation related to propulsion structural longitudinal coupling, Spacecraft Lunar Adapter (SLA) area structural integrity and certain malfunctions of the J-2 engines in the S-II and S-IVB stages. The spacecraft flew the planned trajectory, but preplanned high velocity reentry conditions were not achieved. A majority of the mission objectives for Apollo 6 were accomplished.

The Apollo 7 mission (first manned Saturn IB) was successfully launched on 11 October 1968. This was the fifth and last planned Apollo mission utilizing Saturn IB vehicles (AS-205). The mission provided for the first manned orbital tests of the Block II Command and Service Module. All primary mission objectives were successfully accomplished. In addition, all planned Detailed Test Objectives, plus three that were not originally scheduled, were satisfactorily accomplished.

The Apollo 8 mission will provide the first manned flight of the Saturn V and the first Apollo mission in the lunar environment.

THE APOLLO 8 MISSION

The Apollo 8 mission is the third Saturn V mission and is intended to verify crew, space vehicle, and mission support facilities on a lunar orbit flight plan. The mission is planned for slightly more than six days (147 hours).

The nominal flight plan will closely resemble the manned lunar landing mission planned for later in the Apollo program. Apollo 8 will demonstrate several nominal lunar landing mission activities including translunar injection (TLI), cislunar navigation and communications, lunar orbit insertion (LOI), passive thermal control (PTC), and transearth injection (TEI).

The Apollo 8 launch will be on a selected launch azimuth between 72° and 107°. Launch azimuth selection will be made just prior to launch to effect the desired moon/ spacecraft rendezvous. The ascent-to-earth orbit will include the S-IC and S-II boost phase with the S-IVB orbit insertion burn. The spacecraft will remain attached to the S-IVB in a 100 nautical mile circular parking orbit for approximately two revolutions.

On the second parking orbit, while over the Pacific Ocean, the S-IVB engine will be restarted to insert the CSM and S-IVB into a translunar trajectory. The nominal TLI will provide a "free return" to earth if the deboost into a lunar orbit is not initiated. A second opportunity for TLI will be available on the third revolution if necessary.

Shortly after TLI, the CSM will separate from the S-IVB, transpose, and then move to a safe distance before the S-IVB propellant dump. The S-IVB then executes a retrograde dump of residual propellants and Auxiliary Propulsion System (APS) burn to completion to achieve a "slingshot" effect which reduces the probability that the S-IVB will impact with the CSM, the moon's surface, or return to earth.

The spacecraft will remain in translunar coast (TLC) for approximately three days. During this time navigational sightings, midcourse corrections (MCC), and passive thermal control will be exercised.

Nominally, at approximately 69 hours into the mission, the Service Propulsion System (SPS) will perform the first lunar orbit insertion (LOI-1) maneuver and will place the CSM into a 60 x 170 nautical mile lunar orbit. Following insertion and system checks, the orbit will be circularized at 60 nautical miles by a second SPS maneuver (LOI-2).

The spacecraft will remain in lunar orbit for a total of 10 revolutions (approximately 20 hours) during which time extensive photography and landmark sightings will be made. TV coverage of the lunar surface is planned.

Return to earth will begin when the SPS performs the TEI at approximately 89 hours into the mission. The transearth coast (TEC) will have a duration of approximately 58 hours. A total of three MCC's are allotted during this time for return corridor control. This period will also provide for evaluation of crew activities, navigation sightings, subsystem performance, and mission support facilities.

Approximately two hours prior to entry interface (EI) the Guidance, Navigation and Control System (GNCS) is prepared for an automatic entry. Entry velocity will be approximately 36,220 fps with splashdown planned for the Pacific recovery area. Figure I shows the general mission profile.

NASA MISSION OBJECTIVES

FOR APOLLO 8 PRIMARY OBJECTIVES

Demonstrate crew/space vehicle/mission support facilities performance during a manned Saturn V mission with CSM. Demonstrate performance of nominal and selected backup Lunar Orbit Rendezvous (LOR) mission activities, including:
- Translunar injection.
- CSM navigation, communications, and midcourse corrections.
- CSM consumables assessment and passive thermal control.

George E Mueller Associate Administrator for Manned Space Flight
Sam C. Phillips Lt. General, USAF Apollo Program Director
        Date: 12 Dec. 1968

MISSION PROFILE

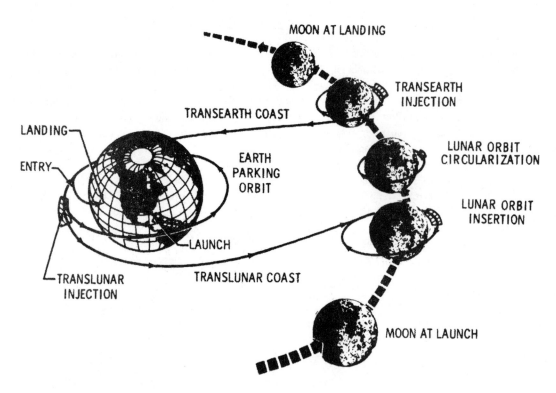

DETAILED TEST OBJECTIVES

The following detail amplifies and defines more explicitly basic tests, measurements, and evaluations which are planned to achieve the primary objectives given previously.

Launch Vehicle

Verify launch vehicle capability for free return translunar injection.

Demonstrate S-IVB restart capability.

Verify J-2 engine modification.

Confirm J-2 engine environment in S-II and S-IVB stages.

Confirm launch vehicle longitudinal oscillation environment during S-IC stage burn period.

Verify modifications incorporated in the S-IC stage to suppress low frequency longitudinal oscillations.

Demonstrate helium heater re-pressurization system operation.

Verify capability to inject S-IVB/IU/LTA-B into a lunar "slingshot" trajectory.

Demonstrate capability to safe S-IVB stage.

Spacecraft

Perform lunar landmark tracking.

Demonstrate ground operational support for a CSM lunar orbit mission.

Obtain data on passive thermal control system.

Perform manual and automatic acquisition, tracking and communications with MSFN using the high-gain. CSM S-band antenna during a lunar mission.

Perform translunar and transearth midcourse corrections.

Perform lunar orbit insertion GNCS controlled SPS burn.

Prepare for translunar injection and monitor the GNCS and LV tank pressure displays during the TLI burn.

Demonstrate SLA panel jettison in a zero-g environment.

Obtain data on the spacecraft dynamic response.

Perform star-earth landmark sightings navigation during translunar and transearth phases.

Perform star-lunar horizon sightings during the translunar and transearth phases.

Perform star-earth horizon sightings during translunar and transearth phases.

Demonstrate CSM passive thermal control modes during the translunar and transearth mission phases.

Perform a GNCS controlled entry from a lunar return.

Perform SPS LOI and TEI burns and monitor the primary and auxiliary gauging systems.

Monitor the GNCS and displays during launch.

Obtain data on the CM crew procedures and timeline for lunar orbit mission activities.

Obtain data an CSM consumables for a CSM lunar orbit mission.

Perform a transearth insertion GNCS controlled SPS burn.

SECONDARY OBJECTIVES

Apollo secondary objectives are established by the development centers to provide additional engineering or scientific data. For Apollo 8 the secondary test objectives are as follows:

Launch Vehicle

Verify the on-board CCS and ground system interface and operation in a deep space environment.

Spacecraft

Communicate with MSFN using the CSM S-band omni antennas at lunar distance.

Demonstrate the performance of the Block II thermal protection system during a manned lunar return entry.

Obtain data to determine the effect of the tower jettison motor, S-II retro and SM RCS exhausts and other sources of contamination on the CSM windows.

Obtain data on the Block II ECS performance during manned lunar return entry conditions.

Obtain IMU performance data in the flight environment.

Perform an IMU alignment and a star pattern visibility check in daylight.

Perform a CSM/S-IVB separation and a CSM transposition on a lunar mission timeline.

Obtain photographs during the transearth, translunar and lunar orbit phases.

SAFETY PLANNING

GENERAL

As a result of the normal developmental cycle, a large number of changes and additions have been made to the Apollo safety systems and procedures. A detailed discussion of changes is not within the scope of this report though some items are covered in later descriptive sections. This section generalizes the procedural or program aspects of safety related improvements.

PRE-LAUNCH PROCEDURES

Procedures for maximum launch complex area safety are provided by "Apollo/Saturn V Ground Safety Plan" K-V-053. Overall safety conditions at the launch pad area are monitored by the Systems Safety Supervisor who has the authority to stop any operation when a condition exists which, in his opinion, is imminently hazardous.

All test procedures are required to contain applicable emergency procedures. Prior to implementation, these test procedures are reviewed by the KSC Safety Office which makes the final determination as to the hazardous classification of all tests. The procedures that are evaluated as hazardous are then compared against a set of checklists to ensure compliance with safety practices.

To ensure maximum readiness of personnel to cope with emergencies, personnel emergency procedure training and practice are given on a regular basis. In addition, training proficiency is reviewed before conducting all hazardous operations. The training program that has been initiated includes the use of fire-fighting equipment, toxic propellant safety, safety locker qualification, first aid and the handling of personnel requiring aid in the event of an emergency. It also includes the development of procedures and training in emergency crew egress. Certification of course completion is mandatory for Launch Complex (LC) 39 personnel.

RANGE SAFETY

The range safety system provides essentially real time space vehicle position, trajectory, and impact prediction information from launch through orbital insertion. The Range Safety Officer (RSO) uses consoles, plotting boards, visual displays, and information from range observers for making his decision on safe or unsafe space vehicle trajectories.

LAUNCH COUNTDOWN AND TURNAROUND CAPABILITY AS-503

Countdown for the Apollo 8 mission will begin with a precount period during which launch vehicle and spacecraft preparations will take place independently until coordinated space vehicle countdown activities begin. Table I shows the significant launch countdown events.

TABLE I
LAUNCH COUNTDOWN SEQUENCE OF EVENTS

COUNTDOWN HRS: MIN: SEC:	EVENT
28:00:00	Start LV and SC countdown activities
24:30:00	S-II power-up
24:00:00	S-IVB power-up
09:00:00	Six hour built-in hold
09:00:00	End of built-in hold; close CM and BPC hatch
08:59:00	Clear pad for LV cryo loading
07:28:00	Start S-IVB LO2 loading
07:04:00	S-IVB LO2 loading complete; start S-II LO2 loading
06:27:00	S-II L02 loading complete; start S-IC LO2 loading
04:57:00	S-IC L02 loading complete
04:54:00	Start S-II LH2 loading
04:11:00	S-II LH2 loading complete; start S-IVB LH2 loading
03:30:00	I hour built-in hold
03:30:00	Flight crew departs MSO
03:28:00	S-IVB LH2 loading complete
03:13:00	Close-out crew on station; start ingress preps
02:40:00	Start flight crew ingress
02:10:00	Flight crew ingress complete
01:40:00	Close SC hatch
01:00:00	Start RP -1 level adjust
00:42:00	Arm LES pyro buses
00:35:00	RP -1 level adjust complete
00:15:00	SC on internal power
00:05:30	Arm S and A Devices
00:03:07	Terminal Count Sequencer (TCS) Start
00:00:17.0	Guidance reference release command
00:00:08.9	S-IC ignition command
00:00:00	Lift-off

SCRUB/TURNAROUND

Scrub/ turnaround times are based upon the amount of work required to return the space vehicle to a safe condition and to complete the recycle activities necessary to resume launch countdown preparation for a subsequent launch attempt. Planning guidelines for the various scrub/turnaround plans are based upon no serial time for repairs or holds, or for systems retesting resulting from repairs; performing tasks necessary to attain launch with the same degree of confidence as for the first launch attempt; and, not requiring unloading of hypergolic propellants and RP-1 from the space vehicle.

TURNAROUND CONDITIONS VS. TIME

Should a hold occur from T-22 minutes to T-16.7 seconds and a recycle to T-22 minutes is required under the conditions stated in the Launch Mission Rules, the applicable recycle operations in the Launch Vehicle and Spacecraft procedures are followed. The decision is then made to hold at this point with the intention of resuming the count, or to scrub and initiate the turnaround procedures. A cutoff after T-16.7 seconds results in a scrub. For a hold prior to T-22 minutes, which results in a scrub, the turnaround procedures are initiated from the point of hold.

Post LV Cryo Load (with Fuel Cell Cryo Reservicing)
        Turnaround time is 67 hours, 30 minutes, consisting of 30 hours, 30 minutes for recycle time and 37 hours for countdown time. Time is based upon scrub occurring between 16.7 seconds and 8.9 seconds during original countdown.

Post LV Cryo Load (No Fuel Cell Cryo Reservicing)
        Turnaround time is 22 hours, 30 minutes, consisting of 13 hours, 30 minutes for recycle time and 9 hours for countdown time.

Pre LV Cryo Load (with Fuel Cell Cryo Reservicing)
        Turnaround time is 61 hours, 15 minutes, consisting of 24 hours 15 minutes for recycle time and 37 hours for countdown time.

Pre LV Cryo Load (No Fuel Cell Cryo Reservicing)
        The capability for approximately a one day hold exists at T-9 hours of the countdown. This capability provides for a launch attempt at the opening of the next launch window.

DETAILED FLIGHT MISSION DESCRIPTION

NOMINAL MISSION

Pre-Launch
        The AS-503 space vehicle for the Apollo 8 mission is planned to be launched from Launch Complex 39, pad A, at Kennedy Space Center, Florida on 21 December 1968. The launch window opens at 0751 EST and closes at 1232 EST on this date. Should holds in the launch countdown or weather require a scrub, there are six days remaining in December during which the mission could be launched. Table 2 shows these days and launch window durations.

TABLE 2
DECEMBER LAUNCH WINDOWS

December Launch Days for Apollo 8	WINDOW (EST)	
	Open	Close
21	0751	1232
22	0926	1405
23	1058	1535
24	1221	1658
25	1352	1820
26	1516	1820
27	1645	1820

        A variable launch azimuth of 72° to 107° capability will be available to assure a launch on time. This is the first Apollo mission which has employed the variable launch azimuth concept. The concept is necessary to compensate for the relative relationship of the earth and moon at launch time.

Launch and Earth Parking Orbit
        The Apollo 8 mission will begin with the boost to orbit by a full burn of the S-IC and S-II stages and a partial burn of the S-IVB stage of the Saturn V launch vehicle. Shortly after S-IC /S-II staging, four camera capsules will be jettisoned from the S-IC. Table 3 shows the mission sequence of events.
        Insertion into the 100-nautical mile earth parking orbit will occur approximately 11 minutes, 32 seconds ground elapsed time (GET) from lift-off. The spacecraft and the S-IVB will remain in this orbit while all systems are checked and readied for the second burn of the S-IVB, the translunar injection (TLI) burn.

TABLE 3
APOLLO 8 SEQUENCE OF EVENTS

Time from Lift-off HR:MIN:SEC:	Event
00:00:00	Lift-off
00:00:11	Roll and Pitch Program initiate
00:00:28	Roll Complete
00:01:17	Maximum Dynamic Pressure
00:02:05	S-IC Center Engine Cutoff
00:02:31	S-IC Outboard Engine Cutoff
00:02:32	S-IC/S-II Separation
00:02:33	S-II Ignition
00:02:55	Camera Capsule Ejection
00:03:00	Second Plane Separation
00:03:07	Launch Escape Tower Jettison Mode I/Mode II Abort Changeover
00:08:40	S-II Cutoff
00:08:41	S-II/S-IVB Separation
00:08:45	S-IVB Ignition
00:09:50	Mode IV Capability Begins
00:10:18	Mode II/Mode III Abort Changeover
00:11:31	Insertion into Earth Parking Orbit
02:50:31	Translunar Injection Ignition
02:55:43	Translunar Injection Cutoff - Translunar Coast Begins
03:09:14	S-IVB/CSM Separation
04:44:54	Begin Maneuver to Slingshot Attitude
05:07:54	LOX Dump Begins
05:12:54	LOX Dump Ends
TLI + 6 Hrs	Midcourse Correction 1
TLI +25 Hrs	Midcourse Correction 2
LOI -22 Hrs	Midcourse Correction 3
LOI - 8 Hrs	Midcourse Correction 4
69:07:30	Lunar Orbit Insertion (LOI-1) Initiation
69:11:36	Lunar Orbit Insertion (LOI-1) Termination
73:30:54	Lunar Orbit Insertion (LOI-2) Initiation
73:31:04	Lunar Orbit Insertion (LOI-2) Termination
89:04:02	Transearth Injection Initiate
89:18:33	Transearth Injection Terminate
TEI +15 Hrs	Midcourse Correction 5
TEI +33 Hrs	Midcourse Correction 6
TEI - 2 Hrs	Midcourse Correction 7
146:35:00	CM/SM Separation
146:50:00	Entry Interface
147:10:00	SPLASHDOWN

Translunar injection and Translunar Coast

TLI will occur during revolution 2 (first opportunity) or 3 (second opportunity) over the Pacific and will last approximately five minutes. A nominal TLI burn will place the spacecraft on a circumlunar, earth-intersecting trajectory. This is called a free return trajectory which means that with no further SPS burns the spacecraft will fly around the moon and safely return to earth.

Within 20 minutes after TLI, the spacecraft will separate from the S-IVB, transpose, and perform the spacecraft/S-IVB evasive maneuver. The S-IVB will then execute the dump of residual propellants plus an APS burn to completion which will impart a differential velocity (Delta-V) to put the S-IVB on a trajectory which is planned to reduce the probability of a S-IVB recontact with the CSM, impacting the moon's surface, or returning to earth.

Periodically during the approximately 66 hours of translunar coast, the spacecraft trajectory will be assessed to determine if a midcourse correction is required. Up to four MCC opportunities have been identified to maintain the free return trajectory and will ensure that the spacecraft will be at least 60 nautical miles above the moon's surface at its point of closest approach.

Lunar Orbit Insertion and Lunar Orbit

When the spacecraft reaches pericynthion (point nearest surface) behind the moon, it performs the

TABLE 4

LUNAR ACTIVITIES

Revolution	Activity
1 & 2	LOI, camera preparation, eat period, COAS ground track determination, control point and pseudo landing site observations, photographs of targets of opportunity and TV transmission.
3 & 4	LOI$_2$, two-hour CMP rest period, landmark training photography, vertical stereo photography and landmark lighting evaluation.
5 & 6	Three-hour CDR rest period, one control point landmark tracking, and a pseudo landing site tracking during each daylight period. Each tracking consists of four marks.
7 & 8	Two-hour LMP rest period, three control point landmark trackings, and a pseudo landing site tracking during each daylight period.
9 & 10	Two-hour CMP rest period, convergent stereo photography an eat period and the TEI maneuver.

NOTE: The IMU is realigned once during each dark period in lunar orbit.

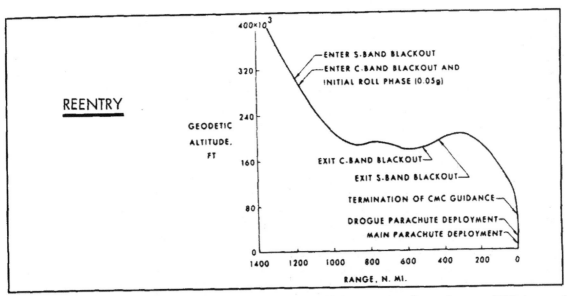

lunar orbit insertion (LOI-I) burn. This is a retrograde Service Propulsion System (SPS) burn of approximately four minutes duration that inserts the spacecraft into a 60 x 170 nautical mile lunar orbit. The spacecraft will remain in this orbit for approximately two revolutions. At third pericynthion, a second SPS burn (LOI-2), lasting approximately 10 seconds, will be performed which will circularize the spacecraft orbit at 60 nautical miles above the lunar surface.

The spacecraft will orbit the moon eight more times for a total of 10 lunar orbits. A 60-nautical mile lunar orbit has a period of approximately two hours; therefore the spacecraft will be in the lunar environment for approximately 20 hours. A schedule of activities while in lunar orbit is shown on Table 4.

Transearch Injection, Transearth Coast, and Entry

While passing behind the moon on the 10th revolution, the spacecraft will perform the transearth injection (TEI) burn. This SPS burn will have a duration of approximately 3.4 minutes and place the spacecraft on a return-to-earth trajectory. During the approximate 58 hours of Transearth coast, the spacecraft will perform up to three course corrections, if required, to ensure that the spacecraft will enter the atmosphere with the proper combination of velocity and flight path angle for safe entry. The spacecraft will enter the atmosphere at 400,000 feet with a velocity of approximately 36,200 feet per second and will land arproximately 1350 nautical miles down range of the entry point. Figure 2. shows entry conditions for Apollo 8. Table 5 lists the entry and landing points for the different days applicable to the December launch windows. Splashdown for the nominal Apollo 8 mission, with a December 21 launch, is planned for approximately six days and three hours after lift-off.

TABLE 5
CM END-OF-MISSION ENTRY AND LANDING POINTS*

Day of Launch	Entry Point Latitude	Longitude	Landing Point Latitude	Longitude
21 Dec	14°42 'N	174° 30'E	4° 5' N	165° 00 'W
22 Dec	5° 35 ' N	173° 50'E	1° 00'5	165° 00' W
23 Dec	1° 20' N	174° 35'E	8° 10'5	165° 00' W
24 Dec	10° 15 'S	172° 15' E	12° 50'5	165° 00' W
25 Dec	18° 55'S	171° 25' E	18° 00'S	165° 00' W
26 Dec	25° 00'S	170° 45'E	22° 10'5	165° 00' W
27 Dec	22° 25'S	170° 55'E	25° 25 '5	165° 00' W

*These points are for a 72° launch azimuth. Other launch azimuths will change the data slightly. TV transmissions from the CM are planned for various phases of the mission (Table 6).

TABLE 6
NOMINAL MISSION TV OPERATION TIME LINE

MSFN Station	Station Time of Acquisition (GET - Hr: Min)	Time (Min)	Remarks
Goldstone	31:15	15	1.Exterior Views (Earth-Moon) 100 mm Lens, 9° FOV
Goldstone	55:15	15	2. Interior Views - Wide Angle Lens; 160° FOV
Madrid	71:35	15	
Goldstone	85:37	15	
Goldstone	104:15	15	
Goldstone	128:00	15	

CONTINGENCY OPERATIONS

If an anomaly occurs after lift-off that would prevent the AS-503 space vehicle from following its nominal flight plan, an abort or an alternate mission will be initiated. An abort would provide only for an acceptable CM/crew recovery while an alternate mission would attempt to achieve some of the mission objectives before providing for an acceptable CM/crew recovery.

ABORTS

Launch Aborts

During launch, the velocity, altitude, atmosphere, and launch configuration change rapidly; therefore, several abort modes, each adapted to a portion of the launch trajectory, are required.

Mode 1 abort procedure is designed for safe recovery of the CM following aborts occurring between Launch Escape System (LES) activation (approximately T minus 30 minutes) and Launch Escape Tower (LET) jettison, approximately 3 minutes GET. The procedure consists of the LET pulling the Command Module (CM) away from the space vehicle and propelling it a safe distance down range. The resulting landing point lies between the launch site and approximately 490 nautical miles down range. The Mode 11 abort would be performed from the time the LET is jettisoned until the full-lift CM landing point is 3200 nautical miles down range, approximately 10 minutes GET. The procedure consists of separating the command and service module (CSM) combination from the launch vehicle, separating the CM from the SM, and then letting the CM free fall to entry. The entry would be a full-lift, or maximum range trajectory, with a landing 400 to 3200 nautical miles downrange on the ground track. Mode III abort can be performed from the time the full-lift landing point range reaches 3200 nautical miles until orbital insertion. The procedure would consist of separating the CSM from the launch vehicle and then, if necessary, performing a retrograde burn with the SPS so that the half-lift landing point is no farther than 3350 nautical miles down range. A half-lift entry would be flown which causes the landing point to be approximately 70 nautical miles south of the nominal ground track between 3000 and 3350 nautical miles down range. The Mode IV abort procedure is an abort to earth parking orbit and could be performed anytime after the SPS has the capability to insert the CSM into orbit. This capability begins at approximately 10 minutes GET. The procedure would consist of separating the CSM from the launch vehicle and, two minutes later, performing a posigrade SPS burn to insert the CSM into earth parking orbit with a perigee of at least 75 nautical miles. The CSM could then remain in earth orbit for an earth orbital alternate mission, or, if necessary, return to earth in the West Atlantic or Central Pacific Ocean after one revolution. This mode of abort is preferred over the Mode III abort and would be used unless an immediate return to earth is necessary during the launch phase. The last phase abort procedure is an Apogee Kick (AK) Mode.

This mode is a variation of the Mode IV wherein the SPS burn to orbit occurs at apogee altitude to raise the perigee to 75 nautical miles. The maneuver is executed whenever the orbital apogee at S-IVB cutoff is favorably situated and the corresponding Mode IV deltaV requirement is greater than 100 feet per second. Like the Mode IV contingency orbit insertion (COI), this maneuver is prime when the capability exists, except for those modified situations where an immediate return to earth is required.

Earth Parking Orbit Aborts

Once the CSM/S-IVB is safely inserted into earth parking orbit, a return-to-earth abort would be performed by separating the CSM from the S-IVB and then utilizing the SFS for a retrograde burn to place the CM on an atmosphere-intersecting trajectory. After entry the crew would fly a guided flight path to a preselected target point if possible.

Aborts Associated with TLI
Ten Minute Abort

If an abort is necessary during TLI the S-IVB would be cut off early and the crew would initiate an onboard calculated retrograde SPS abort burn. The SPS burn would be performed approximately 10 minutes after TLI cutoff and would ensure a safe CM entry. The elapsed time from abort initiation to landing would vary from approximately 25 minutes to 4 hours, depending on the length of the TLI maneuver performed prior to S-IVB cutoff. For aborts initiated during the latter portion of TLI, a second SPS burn could be necessary to correct for dispersed entry conditions. Since this abort would be used only in extreme emergencies with respect to crew survival, the landing point would not be considered in executing the abort. The SM RCS will be used for land avoidance. No meaningful landing point predictions can be made since they will vary with the launch azimuth, the location of the TLI, the duration of the TLI burn prior to cutoff, and execution errors of the abort maneuvers.

Ninety Minute Abort

The ninety minute abort would be applicable after the TLI maneuver has been completed. If any malfunction occurred during the TLI burn, and after careful check, it become apparent that it was necessary to return to earth, an abort procedure specifying an SPS burn at a certain CSM attitude would be transmitted to the crew. The abort would be initiated at approximately TLI cutoff plus 90 minutes. Unlike the previous procedure this abort would be targeted to a preselected landing location called a recovery line. The location of these lines is shown in Figure 8. If possible, the abort would be targeted to the Mid-Pacific recovery line but for some time-critical situations, the abort would be targeted to the Atlantic Ocean line or the East Pacific line. The abort maneuver would be a retrograde SPS burn followed by a midcourse correction, if necessary, performed near apogee to provide the proper CM entry conditions. The elapsed time between abort initiation and landing would normally vary between 11 and 18 hours.

Aborts During Translunar Coast

The abort procedure for this phase of the mission would be similar to the 90 minute abort. If conditions warrant an abort, abort information specifying a combination of SPS burn time and CSM attitude would be sent to the crew. Deep space aborts will be targeted to, in order of priority, (1) the Mid-Pacific line, (2) the Atlantic Ocean line, (3) the East Pacific or West Pacific lines, and (4) the Indian Ocean line. Regardless of the recovery line selected, the landing latitude should remain nearly the same. The minimum elapsed time between abort initiation and CM landing increases with translunar coast flight time. About the time the CSM enters the moon's sphere of gravitational influence, it becomes faster to perform a circumlunar abort rather than returning directly to earth.

Aborts During Lunar Orbit insertion

Aborts following an early shutdown of the SPS during the lunar orbit insertion (LOI) maneuver are divided into two categories, Mode I and Mode III. There is no Mode II procedure on the Apollo 8 Mission because this mode is necessary only for a mission which carries a Lunar Module.

Mode I

The Mode I procedure would be used for aborts following LOI shutdowns from ignition to approximately two minutes into the burn. This procedure would consist of performing a posigrade SPS burn as soon as possible after cutoff to put the CSM back on a return-to-earth trajectory.

Mode III

The Mode III procedure would be used for aborts following LOI shutdowns from approximately two minutes into the burn until nominal cutoff. After two minutes of LOI burn, the CSM will have been inserted into an acceptable lunar orbit. Therefore, the abort procedure would be to let the spacecraft go through one or two lunar revolutions prior to doing a posigrade SPS burn at pericynthion. This would place the CSM on a return-to-earth trajectory targeted to the Mid-Pacific recovery line.

Aborts During Lunar Orbit

Aborts from the lunar orbit would be accomplished by performing the transearth injection burn (TEI) early. The abort would be targeted to the Mid-Pacific recovery line.

Aborts During Transearth Injection

The abort procedures for early cutoff of TEI are the inverse of the LOI abort procedures. That is, for early cutoffs between TEI ignition and approximately two minutes, a Mode III abort would be performed. After this time a Mode I abort would be used. All TEI aborts should result in landings on the Mid-Pacific recovery line.

Aborts During Transearth Coast

From TEI until entry minus 24 hours, the only abort procedure that could be performed is to use the SPS or the SM/reaction control system for a posigrade burn that will decrease the transearth flight time and change the longitude of landing. After entry minus 20 hours, no further abort burns will be performed. This is to ensure that the CM maintains the desired entry velocity and flight path angle combination that will allow a safe entry.

ALTERNATE MISSIONS

Alternate mission plans for Apollo 8 have been prepared for the following contingencies:
1. Failure of the S-IVB late in its first burn which results in an early CSM separation and a subsequent SPS burn to a COI.
2. S-IVB failure prior to TLI or insufficient propellant for restart.
3. Premature or non-nominal TLI termination resulting in an ellipse with associated energy such that a deltaV greater than ~ 3000 fps is required at TLI +3 hours for an SPS MCC.
4. Dispersed trajectories resulting from a malfunction of the S-IVB during TLI.

The alternate missions resulting from the above contingencies are described in the following paragraphs. The descriptions are essentially qualitative since the associated maneuvers will be computed in real time using the real-time auxiliary computing facility (RTACF).

Alternate Mission I

If the S-IVB malfunctions late in its first burn, the CSM has the capability to achieve a low orbit (contingency orbit) using the SPS. And, if the COI requires less than 900 fps, the CSM has the capability to later inject into a 4000 nautical mile apogee ellipse.

If the above conditions exist, the plan is to remain in low earth orbit for approximately 68-72 hours, making MCC's consistent with the lunar timeline. These maneuvers could be used for perigee preservation. In the time period of 68-72 hours GET, the SPS is ignited over US or ETR MSFN stations for a 4200 fps burn to inject the CSM into a 4000 nautical mile apogee ellipse. The injection point is a function of the lighting conditions desired under the apogee for star/landmark navigation sightings.

The CSM remains in the high apogee ellipse for approximately 24 hours. While in the high apogee ellipse, the crew work and rest cycles would be quite similar to that of the lunar mission. The SPS retrograde burn occurs over the US-ETR; the resulting ellipse is approximately 100 x 400 nautical miles. The CSM remains in the low earth orbit for the remainder of the 10-day mission. Remaining SPS propellant can be used for orbit trimming and shaping and deorbit.

In the event the COI deltaV requires more than 900 fps the CSM would not inject into the high ellipse, but remain in low earth orbit. End of mission recovery area lighting conditions are a problem for earth orbit alternates from missions which lift-off early in the first two days of the December launch window. Nodal regression during the 10 days of earth orbit together with the early lift-off results in early morning (pre-sunrise) landings. This condition can be improved in some cases by nodal plane-change maneuvers; this method, however, is somewhat restricted by deltaV capability.

A typical alternate mission I timeline is shown in Table 7 for the 21 December, 72° launch azimuth opportunity.

Alternate Mission II

Alternate mission II is planned for a failure of the S-IVB to restart for the TLI. If, after achieving earth parking orbit (EPO), the S-IVB for any reason cannot be reignited with the CSM attached, the spacecraft can be separated, and the same sequence outlined for the alternate I mission can be followed, except that there is no SPS deltaV limitation as a result of COI. Therefore, the SPS injection to a 4000 nautical mile apogee ellipse would always be performed (assuming the CSM is in a "GO" condition). The same lighting problems for navigation and lighting exists here as in alternate I. Shown in Table 8 is a typical timeline for alternate mission II.

TABLE 7
TYPICAL SPS MANEUVER TIMELINES FOR THE EARTH ORBITAL
ALTERNATE MISSIONS OF THE APOLLO 8
ALTERNATE I MISSION

Mission Time hr: min	Event	Duration min: sec	deltaV fps	Resulting ha/hp, n.mi	MSFN Coverage
0:11	COI	00:56	600	182/100	BDA, insertion ship
9:07	MCC	00:10	110	110/110	HAW
61:00	MCC2	—	—	107/107	—
70:10	SPS injection	05:07	4180	4000/105	TEX, MLA, GBI
93:51	Deboost	03:02	3668	400/105	MLA, GBI
100:03	MCC3	00:15	25	400/90	HAW
165:01	MCC4	00:02	363	200/90	CRO
236:51	Deorbit	00:11	282	190/-10	HAW

All timelines assume 21 Dec. launch, 72° launch azimuth, injection on first opportunity

TABLE 8
TYPICAL SPS MANEUVER TIMELINES FOR THE EARTH ORBITAL
ALTERNATE MISSIONS OF THE APOLLO 8
ALTERNATE 2 MISSION

Mission Time hr: min	Event	Duration min: sec	deltaV fps	Resulting ha/hp, n.mi	MSFN Coverage
09:11	MCC	00:04	45	125/102	HAW
61:00	MCC2	—	—	115/102	—
70:08	SPS injection	05:27	4160	4000/103	TEX, MLA, GBI
93:48	Deboost	03:15	3680	400/103	MLA, GBI
100:01	MCC3	00:15	27	400/90	HAW
165:40	MCC4	00:02	359	200/90	CRO
236:46	Deorbit	00:12	286	190/-10	HAW

All timelines assume 21 Dec. launch, 72° launch azimuth, injection on first opportunity

Alternate Mission 3

Alternate mission 3 contains four subalternates, each depending upon the apogee that can be achieved after early termination of the S-IVB TLI burn.

Alternate 3A

In the event the S-IVB reignites but malfunctions necessitating engine shutdown, and the resulting apogee is less than 4000 nautical miles, the SPS is used to raise the apogee to 4000 nautical miles. The maneuver to achieve the 4000 nautical mile apogee would be performed at the second perigee. The CSM remains in this

orbit for approximately one day (or six orbits), maneuvers to a low earth orbit, and continues a low earth orbit mission. This SPS retrograde maneuver to low earth orbit is performed over the injection ships, roughly 24 hours after TLI. A typical alternate mission 3A timeline is shown in Table 9 for the 21 December, 72° launch azimuth opportunity.

Alternate 3B

If the premature S-IVB shutdown during TLI results in an apogee of 4000 to 25,000 nautical miles, the plan is to perform an SPS phasing maneuver at the first perigee passage to change the orbital period such that at TLI-plus-24-hours the CSM passes through perigee, and such that perigee is located over one or both injection ships, or Hawaii, Canberra, or Carnarvon MSFN stations (depending upon the day of launch, since perigee moves southerly during the launch window). This is done to cover the large SPS burn used to lower the CSM into a 100 x 400 nautical mile low earth orbit. The phasing maneuver could take place anywhere from 6 hours to 16 hours GET, depending upon apogee altitude. This alternate mission sequence is to perform a phasing maneuver at first perigee pass (3 to 14 hours after TLI), coast in the phasing ellipse until approximately TLI-plus-24-hours, maneuver into the 100 x 400 nautical mile orbit, and proceed with a low earth orbit mission (10 days). There is ample time (20 to 25 hours) spent in the high ellipse to perform navigation exercises; however, in some regions of the launch window the landmark lighting conditions underneath the apogee resulting from the TLI premature cutoff are rather poor. A typical timeline for alternate mission 3B is shown in Table 9.

Alternate 3C

If the TLI burn results in an apogee of 25,000 to 60,000 nautical miles the procedure is to perform a retrograde phasing maneuver at the first perigee to alter the orbit period such that a later perigee occurs over a selected Pacific recovery area.

TABLE 9
TYPICAL SPS MANEUVER TIMELINES FOR THE EARTH ORBITAL
ALTERNATE MISSIONS OF THE APOLLO 8   ALTERNATE 3 MISSION

Mission Time hr: min	Event	Duration min: sec	deltaV fps	Resulting ha/hp, n.mi	MSFN Coverage
Alternate 3A					
02:51	TLI c/o	—	—	2000/104	Injection ship,
05:02	Boost to high apogee	02:24	1620	4000/106	Injection ship, GWM
19:56	Deboost	04:13	3679	400/106	TEX, MLA, GBI
63:02	MCC	00:19	29	400/90	TEX, MLA
100:15	MCC2	00:02	337	200/90	CRO
167:00	MCC3	—	—	200/90	—
236:35	Deorbit	00:15	283	188/-10	HAW
Alternate 3B					
02:52	TLI c/o	—	—	23,000/110	Injection ship,
15:37	Phasing	00:07	69	22,100/110	None
27:48	Deboost	08:41	7810	400/110	Injection ship
63:21	MCC1	00:15	35	400/90	MLA, GBI
97:43	MCC2	00:02	336	200/90	CRO
167:00	MCC3	—	—	200/90	—
236:09	Deorbit	00:12	292	196/-11	HAW
Alternate 3C					
02:52	TLI c/o	—	—	50,000/112	Injection ship,
35:28	Phasing	01:00	638	29,900/111	None (Perigee)
70:29	Deboost to semisynchronous	00:46	518	21,800/ 110	Injection ship
186:15	MCC1	—	—	— —	—
236:11	Deorbit	00:04	50	21,800/25	CRO, TAN, CNB

All time lines assume a 21 Dec. launch, 72° launch azimuth, injection first opportunity.

At this perigee another SPS maneuver lowers apogee to approximately 22,000 nautical miles. The resulting ellipse is a semisynchronous (12-hour period) orbit whose perigees occur over the some two points in the Pacific and Atlantic, once each per day. The spacecraft would then remain in this ellipse for 10 days and deorbit from the semisynchronous orbit into the Pacific prime recovery area. Contingency deorbit can be performed from this ellipse at all true anomalies except a relatively small band about perigee (about 25°). This procedure provides two deorbit opportunities per day to stationary locations, one each in the Atlantic and Pacific oceans.

If may be advantageous to make the final orbit slightly less than 12 hours in period to allow the perigee to advance slowly eastward toward the prime area throughout the mission. The recovery ship could move westward, and therefore reduce the amount of required perigee progression. The disadvantage in allowing the Pacific perigee point to progress toward the prime recovery area is that the Atlantic perigee moves onto land. The remainder of the 10 days is spent in this semisynchronous orbit, allowing ample opportunity for navigation exercises. However, for some portions of the launch window, the landmark lighting conditions under the TLI-established apogee position are somewhat marginal.

A typical timeline for alternate mission 3C is shown in Table 9.

Alternate 3D

If the resulting apogee altitude is greater then 60,000 nautical miles, the SPS deltaV required at TLI-plus-3-hours to complete the injection is approximately 3000 fps. At this point, the decision would be made whether or not to perform this maneuver and continue the lunar mission. The TLI-plus-3-hours MCC will always be computed by the RTCC, and if the required deltaV is less than approximately 3000 fps, the MCC would be performed. This alternate could result in a circumlunar flyby or lunar orbit depending upon deltaV available.

Alternate Mission 4

This alternate mission was planned for the contingencies resulting from dispersed trajectories resulting from a S-IVB malfunction during TLI. Planning was based upon an attained apogee of greater than 80,000 nautical miles. In this situation, either a nominal lunar orbit mission of a lunar flyby may be flown. The alternative is dependent upon the deltaV requirements.

There are many variables which could affect this alternate mission; therefore, the RTCC has been programmed for a number of possibilities which can be computed in real time. In general, the real-time decisions will be associated with evaluation and execution of MCC 's.

GO/NO-GO RULES

Go/no-go rules have been established for various plateaus or major phases of the Apollo 8 mission. These rules are presented on Figure 3.

SPACE VEHICLE DESCRIPTION

The Apollo 8 Mission will be performed by on Apollo Saturn V space vehicle designated AS-503, which consists of a three stage Saturn V Launch Vehicle (SA-503) and an Apollo Block 11 Spacecraft (CSM 103). A complete description of the space vehicle and its subsystems is included in the Mission Operation Report Supplement. The following is a brief description of the various stages.

The Saturn V Launch Vehicle consists of three propulsion stages and an Instrument Unit (IU).

The Apollo Spacecraft payload for Apollo 8 consists of a Launch Escape Assembly (LEA), Block 11 Command Service Module, a Spacecraft LM Adapter, and a Lunar Module Test Article.

A list of current weights for the space vehicle is contained in Table 10.

TABLE 10
APOLLO 8 — WEIGHT SUMMARY (All Weights in Pounds)

SEPARATION STAGE/MODULE	INERT WEIGHT	TOTAL EXPENDABLES	TOTAL WEIGHT	FINAL WEIGHT
S-IC Stage	305,650	4,490,140	4,795,790	380,440
S-IC/S-II Interstage	12,610	—	12,610	—
S-II Stage	88,600	949,220	1,037,820	104,280
S-II/S-IB Interstage	8,760	—	8,760	—
S- IV B Stage	26,000	237,640	263,640	29,340
Instrument Unit	4,880	—	4,880	—
Launch Vehicle At Ignition			6,123,500	
SC/LM Adapter	4,060	—	4,060	—
LM Test Article B	19,900	—	19,900	—
Service Module	10,670	40,580	51,250	21,440
Command Module	12,160	—	12,160	11,030
Launch Escape System	8,890	—	8,890	Splshdwn
Spacecraft At Ignition			96,260	
Space Vehicle at Ignition			6,219,760	
S-IC Thrust Buildup			-85,880	
Space Vehicle at Lift-off			6,133,880	
Space Vehicle at Orbit Insertion			284,390	
Space Vehicle at Translunar Injection			121,720	
Spacecraft at Post-TLI Separation			63,410	

LAUNCH VEHICLE DESCRIPTION

First Stage (S-IC)

The S-IC is powered by five F-1 rocket engines each developing approximately 1,522,000 pounds of thrust at sea level and building up to 1.7 million pounds before cutoff. One engine, mounted on the vehicle longitudinal centerline, is fixed; the remaining four engines, mounted in a square pattern about the center line, are gimbaled for thrust vector control by signals from the control system housed in the IU. The F-1 engines utilize LOX (liquid oxygen) and RP-1 (kerosene) as propellants.

Four cameras are mounted in jettisonable capsules on the S-IC forward skirt. Two are arranged to provide coverage of S-IC/S-II separation and the remaining two look into the LOX tank through fiber optics to record behavior during tank depletion. All four capsules are jettisoned after S-IC/S-II separation. Two TV cameras are mounted to provide real time visual monitoring of F-1 engine operation during flight.

Second Stage (S-II)

The S-II is powered by five J-2 rocket engines each developing approximately 200,000 pounds of thrust in a vacuum. One engine, mounted on the vehicle longitudinal centerline, is fixed; the remaining four engines, mounted in square pattern about the center line, are gimbaled for thrust vector control by signals from the control system housed in the IU. The J-2 engines utilize LOX and LH2 (liquid hydrogen) as propellants.

Third Stage (S-IVB)

The S-IVB consists of a cylindrical mainstage powered by a single J-2 engine which is a high performance, multiple-start engine developing approximately 200,000 pounds of thrust in a vacuum. The engine is gimbaled for thrust vector control in pitch and yaw. Roll control is provided by the APS modules containing motors to provide roll control during mainstage operations and yaw and roll control during non-propulsive orbital flight.

Instrument Unit (IU)

The IU contains the following: Electrical system, self-contained and battery powered; Environmental Control, provides thermal conditioning for the electrical components and guidance systems contained in the assembly; Guidance and Control, used in solving guidance equations and controlling the attitude of the vehicle; Measuring and Telemetry, monitors and transmits flight parameters and vehicle operation information to ground stations; Radio Frequency, provides for tracking and command signals.

SPACECRAFT DESCRIPTION

The Apollo 8 spacecraft consists of a Command Module (CM 103), Service Module (SM 103), a Spacecraft Lunar Module Adapter (SLA-11), Lunar Module Test Article-B (LTA-B), and a Launch Escape Assembly (LEA).

Command Module (CM)

CM 103 is a Block II Command Module which contains automatic and manual equipment to control and monitor the spacecraft systems as well as communications equipment and systems to provide for safety and comfort of the flight crew. The primary structure is encompassed by three heat shields forming a truncated, conic structure. The CM consists of a forward compartment, a crew compartment, and an aft compartment (Figure 4).

Service Module (SM)

The Service Module is a cylindrical structure which contains systems to supplement those in the CM. The SM also contains the Service Propulsion System (SPS) which can provide 20,500 pounds of thrust in a vacuum.

Common Command and Service Module Systems

There are a number of systems which are common to the CM and SM.

GO/NO-GO RULE FOR CRITICAL PHASES OF THE APOLLO 8 MISSION

PRINCIPAL GO/NO-GO RULES FOR LAUNCH PHASE

- LAUNCH PHASE CONTINUED TO INSERTION IF AT ALL POSSIBLE SINCE IT IS SAFER TO REENTER FROM EARTH ORBIT THAN ATTEMPT ABORT DURING LAUNCH

- LAUNCH ABORTED FOR:

  - VIOLATION OF AUTO/MANUAL EDS LIMITS
  - TWO ENGINE OUT ON S-II (TIME DEPENDENT)
  - FAILURE OF SECOND PLANE SEPARATION
  - S-IVB LOSS OF THRUST (TIME DEPENDENT)
  - VIOLATION OF TRAJECTORY LIMIT LINES
  - LOSS OF CABIN PRESSURE AND O_2 MANIFOLD LEAK
  - LOSS OF THREE FUEL CELLS AND ONE BATTERY
  - UNCONTROLLABLE SHORTED MAIN BUS
  - LOSS OF BOTH AC BUSES DURING MODE I OR MODE II
  - SUSTAINED LEAK OR LOSS OF H_2 PRESSURE IN BOTH CM-RCS SYSTEMS (MODE I ONLY)

PRINCIPAL GO/NO-GO RULES FOR TLI PHASE

- TLI MANEUVER ATTEMPTED IF AT ALL POSSIBLE IF S-IVB CONSUMABLES ARE SUFFICIENT FOR GUIDED CUTOFF, IF NO S-IVB PROBLEMS ARE PRESENT WHICH WOULD RESULT IN UNSAFE RESTART, AND IF CSM HAS TOTAL SYSTEMS CAPABILITY WITH REDUNDANCY

- TLI INHIBITED FOR:

  - FAILURES IN S-IVB GUIDANCE AND CONTROL SYSTEM
  - FAILURES IN S-IVB PROPULSION SYSTEM OR INSUFFICIENT CONSUMABLES
  - LOSS OF CABIN INTEGRITY, LEAK, FIRE, OR SMOKE IN CABIN
  - LOSS OF SURGE TANK AND REPRESS PACK OR ONE CRYO TANK
  - LOSS OF ONE MAIN O_2 REGULATOR OR LEAK IN O_2 MANIFOLD
  - TOTAL OR PARTIAL LOSS OF PRIMARY OR SECONDARY COOLANT LOOP
  - FAILURE OF BOTH H_2O ACCUMULATORS OR LOSS OF POTABLE OR WASTE H_2O TANK
  - LOSS OF ONE FUEL CELL, ONE ENTRY BATTERY, OR TWO INVERTERS
  - LOSS OF ONE BATTERY, MAIN, AC. OR BATTERY RELAY BUS
  - LOSS OF TWO-WAY COMMUNICATIONS OR GO/NO-GO INSTRUMENTATION
  - PREMATURE ACTIVATION OF SLA-C OR DROGUE CHUTE DEPLOY
  - LOSS OF BOTH BMAGS IN EITHER ROLL, PITCH, OR YAW OR BOTH FDAI'S
  - LOSS OF G & N (CMC, NAV. BRKT, ISS, OR OSS)
  - VIOLATION OF SPS PRESSURE LIMITS
  - LOSS OF BOTH GN_2 TANK PRESSURES
  - ARMING OF CM-RCS SYSTEMS OR LOSS OF ONE SM-RCS QUAD OR CM-RCS SYSTEM

PRINCIPAL GO/NO-GO RULES FOR LOI PHASE

- LOI INHIBITED WITH ENTRY INTO NEXT BEST PTP OR CIRCUMLUNAR FLIGHT ATTEMPTED FOR FAILURES RESULTING IN LOSS OF CAPABILITY OR REDUNDANCY IN THOSE SYSTEMS REQUIRED FOR SAFE RETURN TO EARTH. ABORT METHOD DEPENDENT ON EFFECT OF FAILURE AND PREVAILING CONDITIONS.

  - LOSS OF CABIN INTEGRITY, GLYCOL LEAK, FIRE, OR SMOKE IN CABIN
  - LEAK IN O_2 MANIFOLD OR FAILURE OF MAIN O_2 REGULATOR
  - TOTAL OR PARTIAL LOSS OF PRIMARY OR SECONDARY COOLANT LOOP
  - FAILURE OF BOTH H_2O ACCUMULATORS OR LOSS OF POTABLE OR WASTE H_2O TANK
  - LOSS OF ONE FUEL CELL, ONE ENTRY BATTERY, OR TWO INVERTERS
  - LOSS OF ANY CRYO TANK, OR SURGE TANK AND REPRESS PACK
  - LOSS OF ONE BATTERY, MAIN, AC, OR BATTERY RELAY BUS
  - LOSS OF TWO-WAY COMMUNICATIONS OR GO/NO-GO INSTRUMENTATION
  - PREMATURE ACTIVATION OF SMJC OR DROGUE CHUTE DEPLOY
  - LOSS OF SMAGS IN EITHER ROLL, PITCH, OR YAW OR BOTH FDAI'S
  - LOSS OF EITHER TVC LOOP IN EITHER PITCH OR YAW
  - LOSS OF AUTO ATTITUDE CONTROL IN PITCH AND YAW
  - LOSS OF G & N (CMC, NAV, DSKY, ISS, OR OSS)
  - VIOLATION OF SPS ENGINE PRESSURE OR TEMPERATURE LIMITS
  - FAILURE OF ONE BANK OR BALL VALVES OR GROUND AT ONE SPS SOL DRIVER OUTPUT
  - LOSS OF ONE QUAD, ONE PITCH, ONE YAW, OR TWO ROLL THRUSTERS
  - ARMING OF CM-RCS SYSTEMS OR LOSS OF ONE SM-RCS QUAD OR CM-RCS SYSTEM

PRINCIPAL GO/NO-GO RULES FOR LUNAR ORBIT PHASE

- TRANSEARTH INJECTION PERFORMED AT NEXT BEST OPPORTUNITY FOR FAILURES RESULTING IN DEGRADED SYSTEMS CAPABILITY THAT WOULD AFFECT ABILITY TO PERFORM TEI MANEUVER OR FAILURES IN ECS AFFECTING LIFE SUPPORT

  - LOSS OF CABIN INTEGRITY GLYCOL LEAK, FIRE, OR SMOKE IN CABIN
  - LEAK IN O_2 MANIFOLD
  - LOSS OF PRIMARY COOLANT LOOP
  - LOSS OF POTABLE OR WASTE H_2O TANK
  - LOSS OF ONE FUEL CELL, TWO BATTERIES, OR TWO INVERTERS
  - LOSS OF ONE MAIN, AC, OR BATTERY RELAY BUS
  - PREMATURE ACTIVATION OF SMJC OR DROGUE CHUTE DEPLOY
  - LOSS OF BOTH SMAGS IN EITHER ROLL, PITCH OR YAW OR BOTH FDAI'S
  - LOSS OF DIRECT RCS CONTROL FROM BOTH SMC'S
  - LOSS OF AUTO ATTITUDE CONTROL IN PITCH AND YAW
  - LOSS OF ONE QUAD, ONE PITCH, ONE YAW, OR TWO ROLL THRUSTERS
  - ARMING OF CM-RCS SYSTEMS OR LOSS OF ONE SM-RCS QUAD OR CM-RCS SYSTEM
  - LOSS OF G & N (CMC, NAV, DSKY, ISS, OR OSS)
  - DECAY IN SPS SOURCE, OXID, OR FUEL TANK PRESS, OR EXCESSIVE ω
  - VIOLATION OF SPS ENGINE PRESSURE OR TEMPERATURE LIMITS

Fig. 3

Guidance and Navigation (G&N) System

Measures spacecraft attitude and velocity, determines trajectory, controls spacecraft attitude, controls the thrust vector of the SPS engine, and provides abort information and display data.

Stabilization and Control System (SCS)

Provides control and monitoring of the spacecraft attitude, backup control of the thrust vector of the SPS engine and a backup inertial reference.

Reaction Control System (RCS)

Provides thrust for attitude maneuvers of the spacecraft in response to automatic control signals from the SCS in conjunction with the G&N system.

Electrical Power System (EPS)

Supplies all electrical power required by the CSM. The primary power source consists of three fuel cells which are the prime spacecraft power from lift-off through CM/SM separation. Five batteries - three for entry and post-landing and two for pyrotechnic uses - are located in the CM.

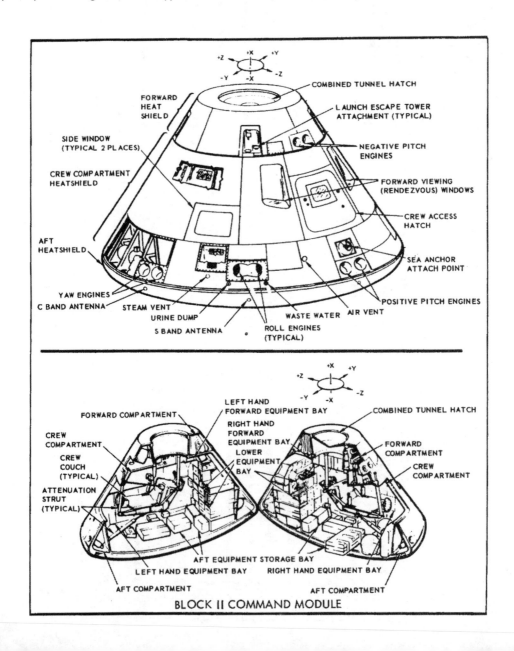

BLOCK II COMMAND MODULE

Preflight crew shots. Anders, Lovell and Borman in the Command Module (Top).
With the Saturn V in the background at KSC (Bottom Right)

The Saturn V Booster (above & bottom left)
The Saturn IVB with the LTA after staging (bottom right)

Earth from 3,500 miles showing Florida and the Bahamas.

Lovell, Anders and Borman on board the Command Module

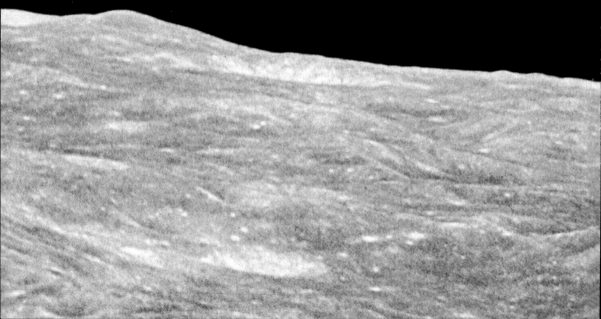

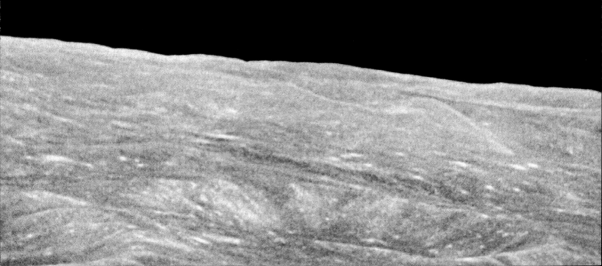

One of the hundreds of amazing lunar surface photographs returned by Apollo 8. This particular shot surprised the crew who had not expected the terracing effect on the crater wall.

From 20,000 miles out the Earth showing South America (left center).

A tired but elated Jim Lovell waits for his turn to be hoisted into the recovery helicopter.

The USS Yorktown recovery team secure the Apollo 8 command module.

Back on board USS Yorktown (above & below)
Taking a call from President Johnson (right)

The crew with a huge cake aboard
USS Yorktown. (above)

Back in the USA with President
Lyndon Johnson (left)

Environmental Control System (ECS)

Provides a controlled cabin environment and dispersion of CM equipment heat loads.

Telecommunications (T/C) System

Provides for the acquisition, processing, storage, transmission and reception of telemetry, tracking, and ranging data between the spacecraft and ground stations.

Sequential (SEQ) Systems

Major subsystems are the sequential events control system (SECS), emergency detection system (EDS), launch escape system (LES), and earth landing system (ELS). The systems interface with the RCS or SPS during an abort.

Spacecraft LM Adapter (SLA)

SLA-11 connects the CSM and the IU and houses the LTA-B. It is a truncated conic structure with a cone element length of 29 feet. The upper section is made up of four 21-foot high panels which swing open at the top and are jettisoned away from the spacecraft by springs attached to the lower fixed panels.

Lunar Module Test Article (LTA-B)

LTA-B (Figure 5) is a steel cylindrical test article instrumented to provide "g" loading values. The "g" loadings are measured by six accelerometers attached to this structure LTA-B is attached to the SLA by four aluminum struts and remains attached throughout the mission.

Launch Escape Assembly (LEA)

The LEA provides the means for separating the CM from the launch vehicle during pad or sub orbital aborts. This assembly consists, in main, of the launch escape tower, launch escape motor, and tower jettison motor. Both motors utilize solid propellants. Redundant analysis and tests have been performed to maximize confidence in this emergency system.

Configuration Differences

The space vehicle for Apollo 8 varies in its configuration from that flown an Apollo 7 and those to be flown on subsequent Apollo missions. These differences are the result of the normal growth, planned changes, and experience gained on previous missions. Figure 6 shows the major configuration differences between AS-205 (CSM 101) and AS-503 (CSM 103).

HUMAN SYSTEM PROVISIONS

The major human system provisions included for the Apollo 8 mission are: Space Suits, Bio-instrumentation System, Medical Provisions, Crew Personal Hygiene, Crew Meals, Sleeping Accommodations, Oxygen Masks, and Survival Equipment. These systems provisions are described in detail in the Mission Operation Report Supplement.

As a result of difficulties with the biomedical harnesses used in Apollo 7, sturdier biomedical harnesses will be worn by the Apollo 8 crew members and a complete spare harness will be carried in stowage. In addition, medication carried on board has been increased as a result of Apollo 7 experience. Other configuration differences, as they apply to human systems, are noted on Figure 6.

LAUNCH COMPLEX

The AS-503 space vehicle (SV) will be launched from Launch Complex 39 at Kennedy Space Center Florida. LC 39 was designed and built to the mobile concept wherein the space vehicle is checked out in an enclosed building before being moved to the pad for final preparations and launch.

The major components of LC 39 include the Vehicle Assembly Building (VAB), the Launch Control Center (LCC) the Mobile Launcher (ML) the Crawler Transporter (C/T) the Mobile Service Structure (MSS) and the Launch Pad.

The LCC is a permanent four story structure located adjacent to the VAB and serves as the focal point for monitoring and controlling vehicle checkout and launch activities for all Saturn V launches.

The ground floor of the structure is devoted to service and support functions. Telemetry equipment occupies the second floor and the third floor is divided into firing rooms, computer rooms and offices. Firing room I will be used for Apollo 8.

The AS-503 space vehicle was received at KSC and assembly and initial overall checkout was performed in the VAB on the mobile launcher. Roll out occurred on 9 October 1968. Transportation to the pad of the assembled space vehicle and ML is provided by the crawler transporter (C/T) which also moves the MSS to the pad after the ML and SV have been secured. The MSS provides 360-degree access to the space vehicle at the launch pad by means of five vertically adjustable elevator serviced enclosed platforms. The MSS is removed to its park position prior to launch.

The emergency egress route system at LC 39 is made up of three major components: the high speed elevators slide tube and slide wire. The primary route for egress from the CM is via the elevators and if necessary through the slide tube which exits into an underground blast room. Apollo 8 is the first mission to employ the slide wire on LC 39. This system is attached to the ML 409 feet above ground and extends approximately 2500 feet west of the ML where it is attached to a 30-foot tower.

Refer to the Mission Operation Report Supplement for a more thorough description of LC 39.

MISSION SUPPORT

Mission support is provided by the Launch Control Center (LCC) the Mission Control Center (MCC) the Manned Space Flight Network (MSFN) and the recovery forces. The LCC is essentially concerned with pre-launch checkout countdown

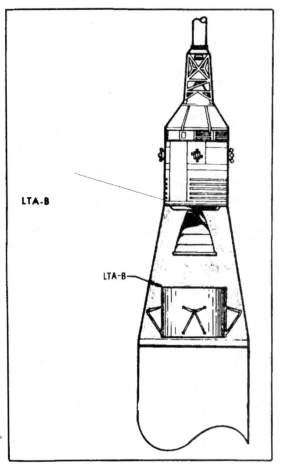

Fig 5

and with launching the space vehicle while MCC located at Houston Texas provides centralized mission control from lift-off through recovery. MCC functions within the framework of a Communications, Command and Telemetry System (CCATS); Real Time Computer Complex (RTCC); Voice Communications System; Display/Control System; and, a Mission Operations Control Room (MOCR). These systems allow the flight control personnel to remain in contact with the spacecraft, receive telemetry and operational data which can be processed by the CCATS and RTCC for verification of a safe mission or compute alternatives. The MOCR is staffed with specialists in all aspects of the mission who provide the flight director with real time evaluations of mission progress.

The MSFN is a worldwide communications network which is controlled by the MCC during Apollo missions. The network is composed of fixed stations (Figure 7) and supplemented by mobile stations (Table 11) which are optimally located within a global band extending from approximately 40° South latitude to 40° North latitude. Station capabilities are summarized in Table 12.

The functions of these stations are to provide tracking, telemetry, command and communications (voice, television) both on an updata link to the spacecraft and on a down data link to the MCC. Connection between these many MSFN stations and the MCC is provided by NASA Communications (NASCOM). Refer to the Mission Operation Report Supplement for more detail on Mission Support.

The Apollo 8 mission will be the first opportunity to acquire mission experience for the Deep Space Instrumentation Facilities (DSIF). There are three stations in this network - Madrid (MAD), Goldstone (GDS), and Honeysuckle Creek (HSK). These stations, each having 85-foot antennas, will be tested with the onboard spacecraft high-gain and omni-antennas to evaluate long range transmissions.

Real-time TV signals from the spacecraft will be available through Madrid and Goldstone to the MCC.

MAJOR CONFIGURATION DIFFERENCES

STRUCTURES

	101	103
Modify Forward Hatch to a Combined Forward Crew Hatch		X
Structural Mod to SM Aft Bulkhead to Assure a 1.4 Factor of Safety for Saturn V		X
Increase CM-SM Tension Tie Thickness		X
Redesign SM/SLA Interface to Install Bolts from Outside		X
Reduce Couch Strut Load/Stroke Criteria and Add Lockouts		X
Addition of 0.03 inch of Cork on SLA		X
Addition of LTA-B		X
Unitized Outboard Couches	X	
Foldable Couches (All Positions)		X

COMMUNICATIONS SUBSYSTEM

	101	103
Conversion of Spacecraft Ground Intercom from Two-Wire to Four-Wire System		X
Add S-band High-Gain Antenna		X
High-Gain Antenna Automatic Reacquisition		X

ENVIRONMENTAL CONTROL SUBSYSTEM

	101	103
Redesign ECS Radiator Flow Proportioning valve		X
Change Material of CO$_2$ Absorber Elements — Stainless — Aluminum	X	
Demand Regulator Capability to — Connect Suit-to-cabin ΔP Transducer to DC Interface connector		X
Relocation of Suit-to-Cabin ΔP Transducer		X
Signal Amplifier Internal Circuit Change		X
Cabin Pressure Relief Valve—Change Stroke to Reduce Variation in Relief Pressure		X

SERVICE PROPULSION SUBSYSTEM (SPS)

	101	103
ID Ball valve to -IE Ball valve		X
Propellant Gaging Bias Correction		X

SPACE SUIT

	101	103
Provide Hard Ring Neck Dam	X	
Incorporate IVCL on EV PGA's		X
Provide Larger Helmet (Borman Only)		X
Relocate Gas Connectors Inward		X
Redesign Knee Convolute Cable Ending		X
IVCL Cross Section Change		X

GUIDANCE AND NAVIGATION

	101	103
Onboard Software –Sundisk to Colossus		X

STABILIZATION CONTROL SUBSYSTEM (SCS)

	101	103
Single Jet Isolation to Provide Separate SCS Power Control — Powered From Bank Switches	X	
— Powered From Separate Switches		X

ORDNANCE

	101	103
SLA Panels (Jettisonable)		X
LM Ejection Thruster		X

INSTRUMENTATION

	101	103
Van Allen Belt Dosimeter		X
Add POGO Instrumentation		X
Nuclear Particle Detection System		X
Delete Two SPS Transfer Line Temperature Measurements and add Two Temperature Measurements to Propellant Utilization Valve (High and Low Bit Rate)		X

DISPLAYS AND CONTROLS

	101	103
IU Up TLM Inhibit Switch on Panel No. 2		X
Redundant Launch Vehicle Attitude Error Display (Hardware change for MSFC only)		X

CREW EQUIPMENT

	101	103
Deletion of Right-Hand Crewman's Right-hand Arm Rest		X

CM – RCS

	101	103
Add Holes to RCS Tank Abrasion PADS to Allow Depressurization of He		X

ENTRY MONITOR SUBSYSTEM

	101	103
Entry Monitor Scroll Patterns Changed From Earth Orbital To Lunar Mission		X

In addition, the 210-foot antenna at Goldstone will be in use for Apollo 8. This antenna will perform a backup function and will be used in a passive mode recording telemetry and voice transmissions.

TABLE 11
MSFN MOBILE FACILITIES APOLLO SHIPS (4 required)

FUNCTION	SUPPORT	LOCATION	NAME
Apollo Insertion Ship	Insertion, abort contingencies	25°N, 49°W	USNS VANGUARD
Apollo injection Ship	Orbital event support	25° N. 155.5°E	USNS REDSTONE
Apollo Injection Ship	Also reentry area support ship	7.5°N, 178°W - I 18°N, 159°E - R	USNS MERCURY
Apollo Reentry Ship	Also supports injection	25°N, 175°W - I 10°N, 177°E - R	USNS HUNTSVILLE

APOLLO AIRCRAFT (5 required)
        A/RIA will support the mission on specified revolutions from assigned Test Support Positions (TSP). In addition, A/RIA will cover reentry (400,000 ft) through crew recovery. A/RIA #1, #2 and #3 will operate in the Pacific Sector and A/RIA #4 and #5 in the Indian Ocean.
        I - Injection position
        R - Reentry position

RECOVERY
        Recovery operations begin with touchdown of the Command Module and are terminated with the retrieval of the Command Module and astronauts. The recovery forces are made up of ships and aircraft deployed in the planned recovery areas (TABLE 13). These areas include the Primary landing areas and abort and contingency landing areas.

TABLE 13
RECOVERY SHIP DEPLOYMENT

Ship	Station	Ship (Figs. 8,9,10,11,12, and 13
SRS	I	USS Guadalcanal (LPH -7)
SRS	2	USS Rankin (AKA -103)
SRS	3	USNS Vanguard (AIS)
SRS	4	USS Chukawan (AO-100)
SRS	5	USS Sandoval (LPA -194)
SRS	6	USS Francis Marion (LPA-249)
PRS		USS Yorktown (CVS- 10)
SRS	7	USS Rupertus (DD 851)
SRS	8	USS Cochrane (DDG -21)
SRS	9	USS Nicholas (DD 449)

        Various safety features are included on the CM to aid the astronauts and recovery team. Among these are a recovery light, sea marker dye, VHF communications, and uprighting bags which are inflated should the CM land in an inverted position.
        The recovery team is trained to provide a fast and safe recovery of the CM. Upon location of the CM, helicopters with swim teams on board are launched to deliver a flotation collar to the CM which will provide flotation capability for a minimum of 48 hours. The crew may be recovered by the helicopters or remain in the CM and be picked up by ship.

TABLE 12

NETWORK CONFIGURATION FOR THE AS-503 MISSION

Facilities	Tracking				USB				TLM					CMD	Data Processing			Comm					Other	
(Systems)	C-band (High Speed)	C-band (Low Speed)	ODOP	Optical	USB	Voice (A/G)	Command	Telemetry	VHF Links	FM Remoting	Mag Tape Recording	Decoms	Displays	CMD Destruct	642B TLM	642B CMD	1218	High Speed Data	Wideband Data	TTY	Voice (SCAMA)	VHF A/G Voice	Video (TV)	SPAN
CIF											X	X						X	X		X		X	
TEL 4									X	X	X													
CNV	X		X	X										X								X		
PAT*	X	X																						
MLA	X	X																						
MIL					X	X	X	X	X	X	X	X	X		X	X	X	X			X	X	X	X
GBI	X*	X*							X		X			X	X									
GBM					X	X	X	X	X	X	X	X			X	X	X	X			X	X	X	X
GTK	X	X												X										
ANG					X	X	X	X	X	X	X	X			X	X	X	X			X	X	X	X
ANT	X	X							X		X													
BDA	X	X			X	X	X	X	X	X	X	X	X	X	X	X	X	X			X	X	X	X
ACN					X	X	X	X	X	X	X	X			X	X	X	X			X	X	X	X
ASC	X	X																						
MAD					X	X	X	X		X	X	X			X	X	X	X			X	X	X	
MADX					X	X	X	X									X				X	X		
CYI		X			X	X	X	X	X	X	X	X			X	X	X	X			X	X	X	X
PRE		X																						
TAN		X							X		X										X	X	X	
CRO	X	X			X	X	X	X	X	X	X	X			X	X	X	X			X	X	X	X
HSK					X	X	X	X		X	X	X			X	X	X	X			X		X	
HSKX					X	X	X	X									X				X			
GWM					X	X	X	X	X	X	X	X			X	X	X	X			X	X	X	X
HAW		X			X	X	X	X	X	X	X	X	X		X	X	X	X			X	X	X	X
CAL	X	X																			X	X	X	
WHS	X	X																			X	X		
GDS					X	X	X	X		X	X	X			X	X	X	X			X		X	
GDSX					X	X	X	X									X				X			
GYM					X	X	X	X	X	X	X	X			X	X	X	X			X	X	X	X
TEX					X	X	X	X	X	X	X	X			X	X	X	X			X	X	X	X
HTV		X			X	X		X			X	X									X	X	X	
RED	X	X			X	X	X	X	X	X	X	X			X	X			X	X	X	X	X	X
VAN	X	X			X	X	X	X	X	X	X	X			X	X				X	X	X	X	X
MER	X	X			X	X	X	X	X	X	X	X			X	X				X	X	X	X	X
ARIA (6)					X	X		X	X		X											X	X	

*Subject to availability.

ORE

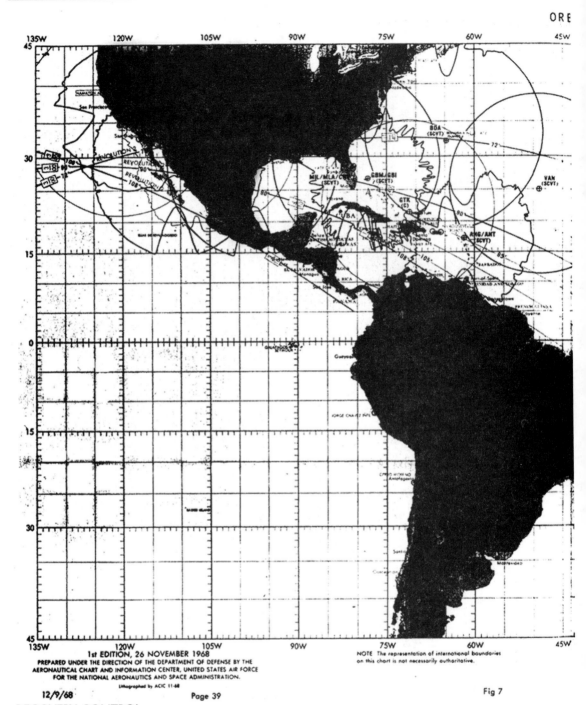

1st EDITION, 26 NOVEMBER 1968
PREPARED UNDER THE DIRECTION OF THE DEPARTMENT OF DEFENSE BY THE
AERONAUTICAL CHART AND INFORMATION CENTER, UNITED STATES AIR FORCE
FOR THE NATIONAL AERONAUTICS AND SPACE ADMINISTRATION.
Lithographed by ACIC 11-68

12/9/68   Page 39   Fig 7

NOTE The representation of international boundaries
on this chart is not necessarily authoritative.

RECOVERY CONTROL

Recovery forces for the Apollo 8 mission will be deployed in both the Atlantic and Pacific Ocean (Figure 8 and Table 13). The recovery will be directed from the Recovery Control Room of the MCC and will be supported by two satellite recovery control centers: the Atlantic Recovery Control Center located at Norfolk, Virginia, and the Pacific Recovery Control Center located at Kunia in the Hawaiian Islands. In addition to the recovery control centers, there will be NASA representatives deployed with recovery forces throughout the worldwide DOD recovery network.

There are five recovery areas which contain the spacecraft landing points following aborts, alternate missions, and the nominal Apollo 8 mission. The areas are: Launch Site Area, Launch Abort Area, Primary Landing Area, Secondary Landing Area and Contingency Landing Areas.

BITAL EVENTS GROUND TRACK

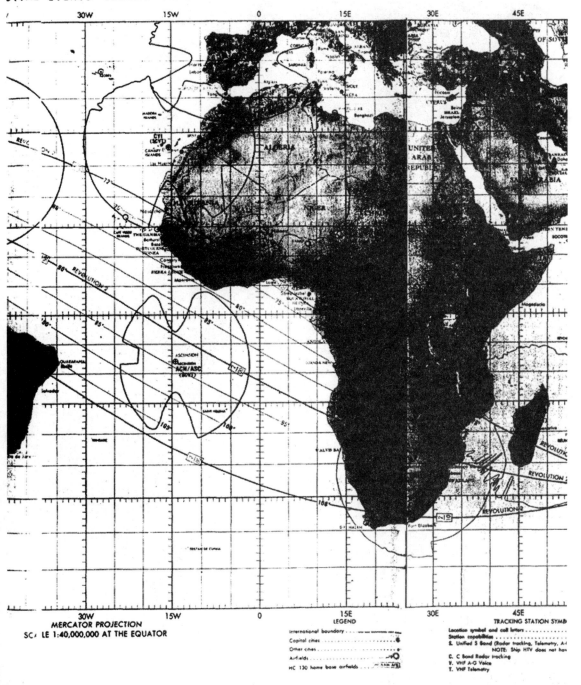

MERCATOR PROJECTION
SC∕ LE 1:40,000,000 AT THE EQUATOR

LEGEND

International boundary . . . .
Capital cities . . . .
Other cities . . . .
Airfields . . . .
HC 130 home base airfields . . . . . .

TRACKING STATION SYMB

Location symbol and call letters . . . .
Station capabilities . . . .
S. Unified S Band (Radar tracking, Telemetry, A∕
         NOTE: Ship HTV does not hav
C. C band Radar tracking
V. VHF A-G Voice
T. VHF Telemetry

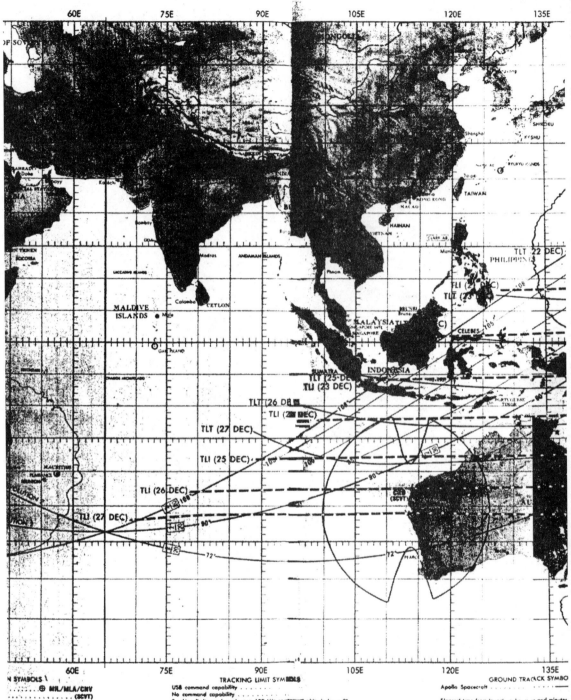

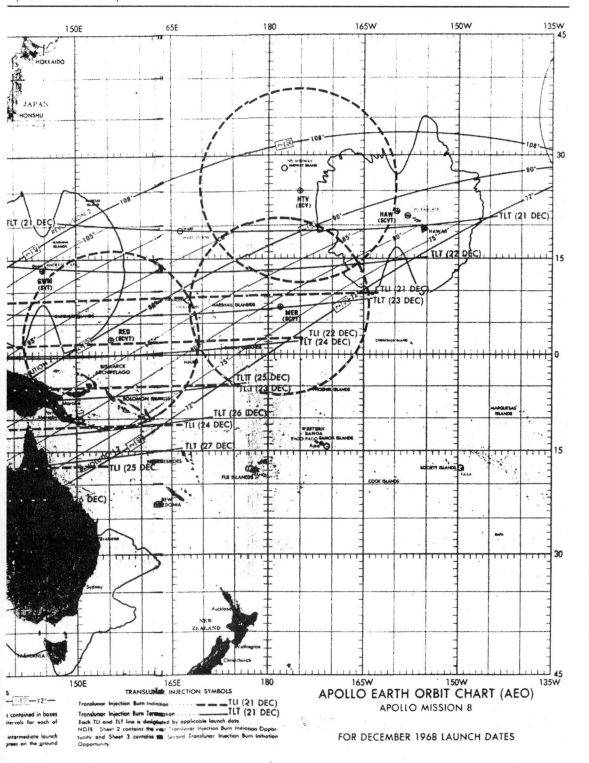

TRANSLUNAR INJECTION SYMBOLS

APOLLO EARTH ORBIT CHART (AEO)
APOLLO MISSION 8

FOR DECEMBER 1968 LAUNCH DATES

Transunar Injection Burn Initiation . . . . — — — TLI (21 DEC)
Translunar Injection Burn Termination . . . . . . —————TLT (21 DEC)
Each TLI and TLT line is designated by applicable launch date.
NOTE: Sheet 2 contains the first Translunar Injection Burn Initiation Opportunity and Sheet 3 contains the Second Translunar Injection Burn Initiation Opportunity.

LAUNCH SITE AREA

The launch site area is that area in which the CM will land following aborts which could occur between launch escape system activation and approximately 90 seconds ground elapsed time (GET).

For a specific wind direction and velocity at the launch site, the locus of possible CM landing points will lie within a relatively narrow corridor in the launch site area. This corridor, based on winds and launch azimuth, will be defined and passed to the launch site recovery forces just prior to launch. The Launch Site Area is illustrated in Figure 9.

LAUNCH ABORT AREA

The launch abort area is that area in which the CM will land following on abort initiated during the launch phase of flight. Since the launch azimuth may vary from 72° to 107°, the launch abort area is designed to include all possible CM landing points following a launch abort from any launch azimuth. Since a landing following a launch abort will be on or near the ground track, the locus of possible CM landing points is a relatively narrow corridor within the launch abort area once the launch azimuth is determined.

The launch abort area (Figure 10) is divided into two sectors, A and B. These sectors are used to differentiate between the recovery force support required in the eastern and western portions of the area. Sector A is all the area in the launch abort area that is within 1200 nautical miles of the launch site. This sector includes a majority of the landing points that occur with a "high g" entry trajectory. Sector B is all the area in the launch abort area that is between 1200 and 3400 nautical miles of the launch site.

Located in the western portion of sector A is the camera capsule landing area (Figure 11). This area, bounded by 335 and 365 downrange lines, contains the predicted camera capsule landing points and the high probability dispersion areas (30 by 15 nautical mile ellipses) around the landing points for any launch azimuth.

PRIMARY LANDING AREA

A primary landing area is an area in which a landing could occur after launch and the probability of such a landing is sufficiently high to warrant a requirement for primary recovery ship (PRS) support. The area is designed to encompass the spacecraft target point and the surrounding dispersion area associated with a high-speed entry from deep space. The areas are used to define recovery support for a CM landing following the nominal mission or an abort initiated any time after the completion of TLI. For the Apollo 8 mission, the Primary landing area is bounded by an ellipse 800 nautical miles long by 300 nautical miles wide. Primary landing areas will be located on or near the Mid-Pacific recovery line, one of five recovery lines designated as general locations where landing areas may be selected (Figure 12).

SECONDARY LANDING AREA

A secondary landing area is an area in which a landing could occur after launch and the probability of such a landing is sufficiently high to warrant a requirement for at least secondary recovery ship (SRS) support. The area is designed to encompass the spacecraft target point and its associated high-probability landing point dispersion.

Deep space secondary landing areas are designed to encompass the target point and dispersion area associated with a high-speed entry from deep space. These will be used to define recovery support for a CM landing from an abort following the completion of TLI. These areas are bounded by an ellipse 800 nautical miles long by 300 nautical miles wide. Deep space secondary landing areas will be selected on or near the Atlantic Ocean recovery line (Figure 13).

CONTINGENCY LANDING AREA

The contingency landing area is all the area outside the launch site, launch abort, and primary and secondary landing areas within which a landing could possibly occur. These contingency areas are shown on Figure 8.

APOLLO 8 RECOVERY FORCE DEPLOYMENT

RECOVERY LINES AND CONTINGENCY AREAS

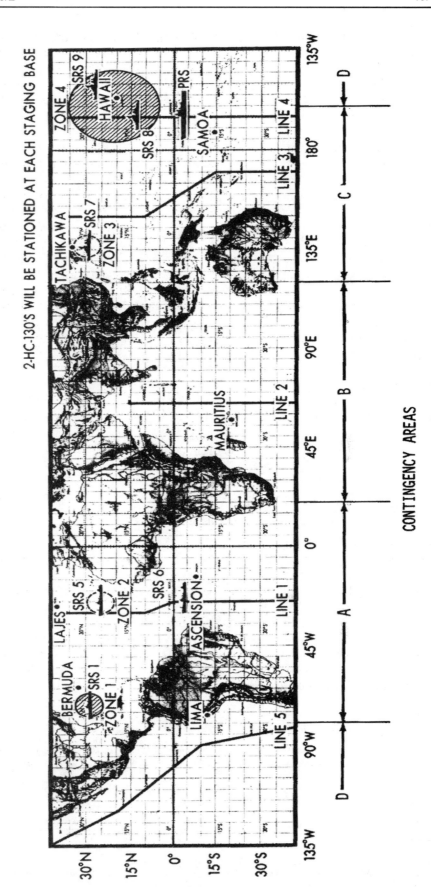

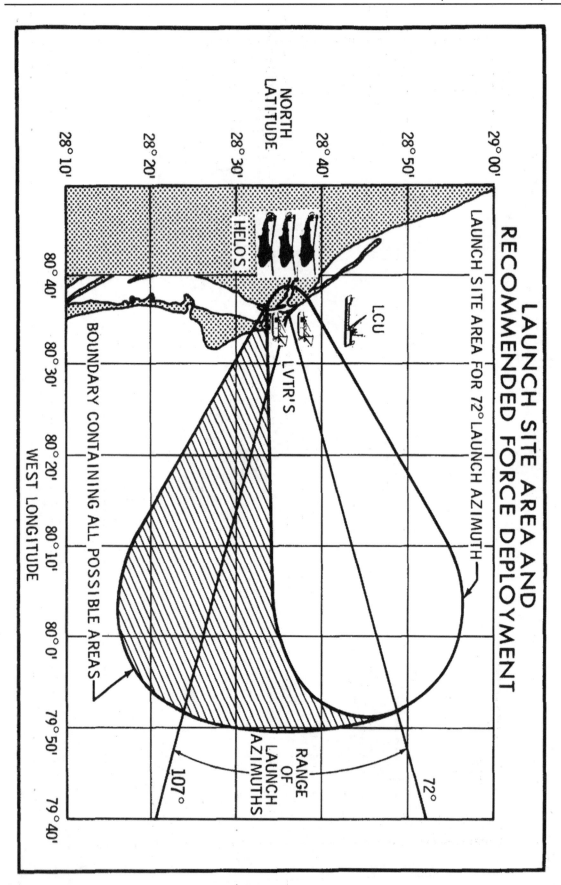

LAUNCH SITE AREA AND
RECOMMENDED FORCE DEPLOYMENT

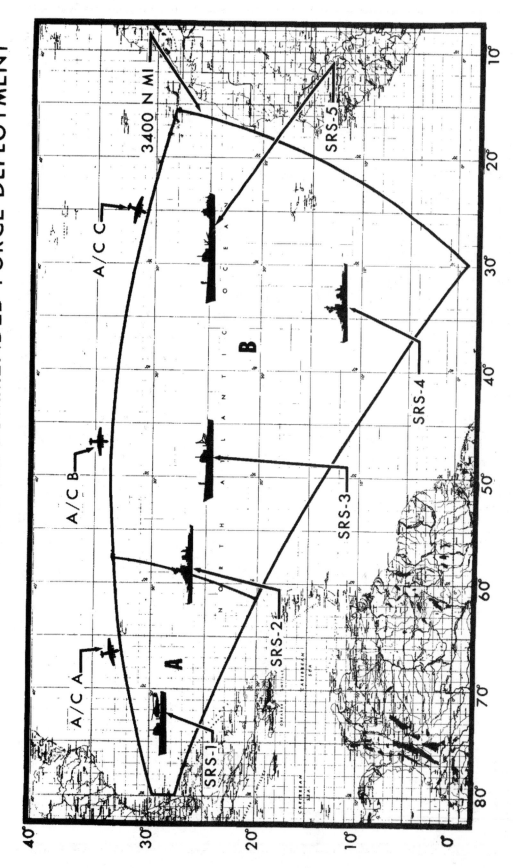

LAUNCH ABORT AREA AND RECOMMENDED FORCE DEPLOYMENT

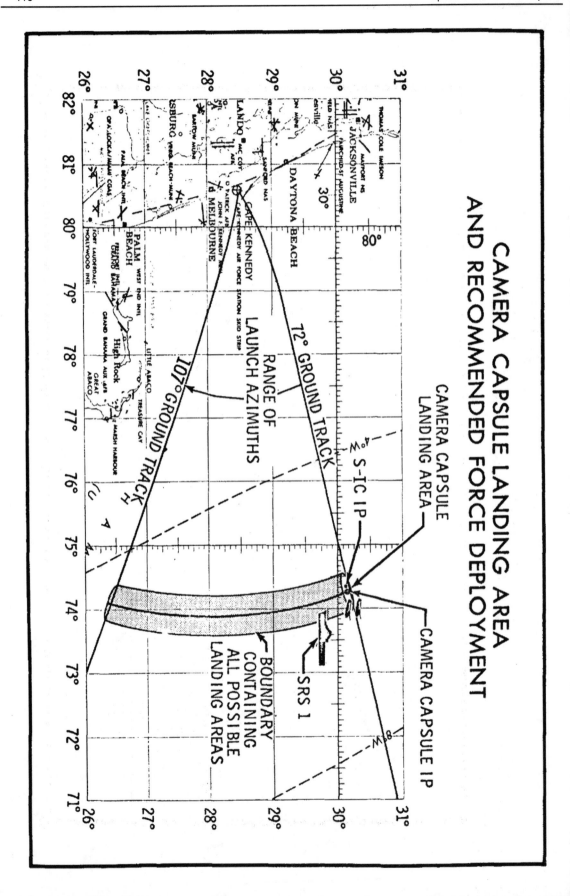

CAMERA CAPSULE LANDING AREA
AND RECOMMENDED FORCE DEPLOYMENT

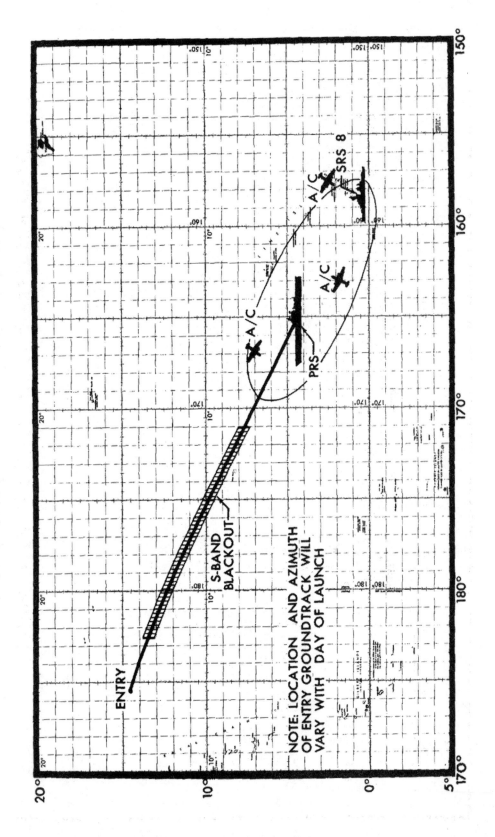

PRIMARY LANDING AREA AND
RECOMMENDED FORCE DEPLOYMENT

NOTE: LOCATION AND AZIMUTH
OF ENTRY GROUNDTRACK WILL
VARY WITH DAY OF LAUNCH

DEEP SPACE SECONDARY LANDING AREA
AND RECOMMENDED FORCE DEPLOYMENT

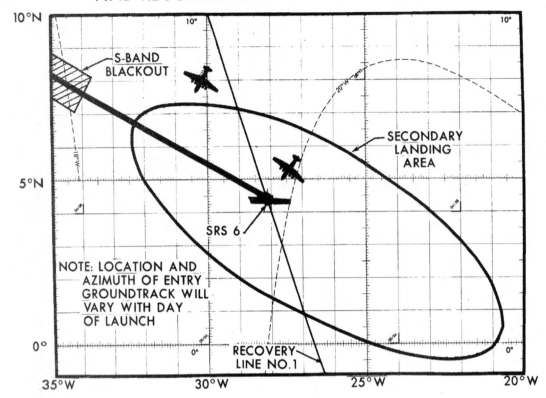

FLIGHT CREW

FLIGHT CREW ASSIGNMENTS

Prime Crew (Figure 14)

Command Pilot (CDR) - Frank Borman, Col., USAF
Command Module Pilot (CMP) - James A. Lovell, Jr., Capt., USN
Lunar Module Pilot (LMP) - William A. Anders, Maj., USAF
Backup Crew (Figure 15)

Command Pilot (CDR)- Neil A. Armstrong (Mr.) Command Module Pilot (CMP) - Edwin E. Aldrin, Jr., Col.,
USAF Lunar Module Pilot (LMP) - Fred W. Haise, Jr., (Mr.)

PRIME CREW BIOGRAPHICAL DATA

Command Pilot (CDR)
NAME: Frank Borman (Colonel, USAF)
DATE OF BIRTH: 14 March 1928
PHYSICAL DESCRIPTION: Height: 5 feet, 10 inches; weight: 163 pounds
EDUCATION: Received a Bachelor of Science degree from the United States Military Academy at West Point
in 1950 and a Master of Science degree in Aeronautical Engineering from the California Institute of
Technology, Pasadena, California, in 1957.
ORGANIZATIONS: Member of the American Institute of Aeronautics and Astronautics and the Society of
Experimental Test Pilots.
SPECIAL HONORS: Awarded the NASA Exceptional Service Medal, Air Force Astronaut Wings, and Air Force

Distinguished Flying Cross: recipient of the 1966 American Astronautical Flight Achievement Award and the 1966 Air Force Association David C. Schilling Flight Trophy; co-recipient of the 1966 Harmon International Aviation Trophy; and recipient of the California Institute of Technology Distinguished Alumni Service Award for 1966.

EXPERIENCE: Borman, an Air Force Colonel, entered the Air Force after graduation from West Point and received his pilot training at Williams Air Force Base, Arizona.

From 1951 to 1956, he was assigned to various fighter squadrons in the United States and the

Figure 14.

James A. Lovell Jr.   William A. Anders   Frank Borman

Philippines. He became an instructor of thermodynamics and fluid mechanics at the Military Academy in 1957 and subsequently attended the USAF Aerospace Research Pilots School from which he graduated in 1960. He remained there as an instructor until 1962.

He has accumulated over 5500 hours flying time, including 4500 hours in jet aircraft.

CURRENT ASSIGNMENT: Colonel Borman was selected as an astronaut by NASA in September 1962. He has performed a variety of special duties, including an assignment as a member of the Apollo 204 Review Board.

As command pilot of the history-making Gemini 7 mission, launched on 4 December 1965, he participated in establishing a number of space "firsts" — among which are the longest manned space flight (330 hours and 35 minutes) and the first rendezvous of two manned maneuverable spacecraft as Gemini 7 was joined in orbit by Gemini 6.

Command Module Pilot (CMP)
NAME: James A. Lovell, Jr. (Captain, USN)
DATE OF BIRTH: 25 March 1928
PHYSICAL DESCRIPTION: Height: 5 feet, 11 inches; weight: 170 pounds.
EDUCATION: Graduated from Juneau High School, Milwaukee, Wisconsin; attended the University of Wisconsin for two years, then received a Bachelor of Science degree from the United States Naval Academy in 1952.
ORGANIZATIONS: Member of the Society of Experimental Test Pilots and the Explorers Club.
SPECIAL HONORS: Awarded two NASA Exceptional Service Medals, the Navy Astronaut Wings, two Navy Distinguished Flying Crosses, and the 1967 FAI Gold Space Medal (Athens, Greece); and co-recipient of the 1966 American Astronautical Society Flight Achievement Award and the Harmon International Aviation Trophy in 1966 and 1967.
EXPERIENCE: Lovell a Navy Captain, received flight training following graduation from Annapolis.

He has had numerous naval aviator assignments including a 4-year tour as a test pilot at the Naval Air Test Center, Patuxent River, Maryland. While there he served as program manager for the F4H weapon system evaluation. A graduate of the Aviation Safety School of the University of Southern California, he also served as a flight instructor and safety officer with Fighter Squadron 101 at the Naval Air Station, Oceana, Virginia.

Of the 4000 hours flying time he has accumulated, more than 3000 hours are in jet aircraft.

CURRENT ASSIGNMENT: Captain Lovell was selected as an astronaut by NASA in September 1962.

On 4 December 1965, he and command pilot Frank Borman were launched on the Gemini 7 mission. The flight lasted 330 hours and 35 minutes.

The Gemini 12 mission, with Lovell and pilot Edwin Aldrin, began on 11 November 1966. This 4-day, 59-revolution flight brought the Gemini Program to a successful close. Major accomplishments of the 94 hour, 35 minute flight included a third-revolution rendezvous with the previously launched Agena (using for the first time backup onboard computations due to a radar failure) a tethered station-keeping exercise; retrieval of a micro-meteorite experiment package from the spacecraft exterior; an evaluation of the use of body restraints specially designed for completing work tasks outside of the spacecraft; and completion of numerous photographic experiments, the highlights of which are the first pictures taken from space of an eclipse of the sun.

Lunar Module Pilot (LMP)
NAME: William A. Anders (Major, USAF)
DATE OF BIRTH: 17 October 1933
PHYSICAL DESCRIPTION: Height: 5 feet, 8 inches; weight: 145 pounds
EDUCATION: Received a Bachelor of Science degree from the United States Naval Academy in 1955 and a Master of Science degree in Nuclear Engineering from the Air Force Institute of Technology at Wright Patterson Air Force Base, Ohio, in 1962.
ORGANIZATIONS: Member of the American Nuclear Society and Tau Beta pi.
SPECIAL HONORS: Awarded the Air Force Commendation Medal.
EXPERIENCE: Anders, an Air Force Major, was commissioned in the Air Force upon graduation from the Naval Academy. After Air Force flight training, he served as a fighter pilot in all-weather interceptor squadrons of the Air Defense Command.

After his graduate training, he served as a nuclear engineer and instructor pilot at the Air Force Weapons Laboratory, Kirtland Air Force Base, New Mexico, where he was responsible for technical management of radiation nuclear power reactor shielding and radiation effects programs.

He has logged more than 3000 hours flying time.

CURRENT ASSIGNMENT: Major Anders was one of the third group of astronauts selected by NASA in October 1963 and has since served as backup pilot for the Gemini 11 mission.

BACKUP CREW BIOGRAPHICAL DATA

Command Pilot (CDR)
NAME: Neil A. Armstrong (Mr.)
DATE OF BIRTH: 5 August 1930
PHYSICAL DESCRIPTION: Height: 5 feet, 11 inches; weight: 165 pounds.
EDUCATION: Attended secondary school in Wapakoneta, Ohio; received a Bachelor of Science degree in Aeronautical Engineering from Purdue University in 1955. Graduate School - University of Southern California.
ORGANIZATIONS: Associate Fellow of the Society of Experimental Test Pilots; Associate Fellow of the American Institute of Aeronautics and Astronautics; and member of the Soaring Society of America.
SPECIAL HONORS: Recipient of the 1962 Institute of Aerospace Sciences Octave Chanute Award; the 1966 AIM Astronautics Award; the NASA Exceptional Service Medal; and the 1962 John J. Montgomery Award.
EXPERIENCE: Armstrong was a naval aviator from 1949 to 1952 and flew 78 combat missions during the Korean action.

He joined NASA's Lewis Research Center in 1955 (then NACA Lewis Flight Propulsion Laboratory) and later transferred to the NASA High Speed Flight Station (now Flight Research Center) at Edwards Air Force Base, California, as an aeronautical research pilot for NACA and NASA. In this capacity, he performed as an X-15 project pilot, flying that aircraft to over 200,000 feet and approximately 4000 miles per hour.

Other flight test work included piloting the X-1 rocket airplane, the F-100, F-101, F-102, F-104, F5D B-47, the paraglider, and others.

He has logged more than 4000 hours flying time.

CURRENT ASSIGNMENT: Mr. Armstrong was selected as an astronaut by NASA in September 1962. He served as backup command pilot for the Gemini 5 and Gemini 11 flights.

As command pilot for the Gemini 8 mission, which was launched on 16 March 1966, he performed the first successful docking of two vehicles in space.

Command Module Pilot (CMP)
NAME: Edwin E. Aldrin, Jr. (Colonel, USAF)
DATE OF BIRTH: 20 January 1930
PHYSICAL DESCRIPTION: Height: 5 feet, 10 inches; weight: 165 pounds.
EDUCATION: Graduated from Montclair High School, Montclair, New Jersey; received a Bachelor of Science degree from the United States Military Academy at West Point, New York, in 1951 and a Doctor of Science degree in Astronautics from the Massachusetts Institute of Technology in 1963; recipient of an Honorary Doctorate of Science degree from Gustavus Adolphus College in 1967.
ORGANIZATIONS: Associate Fellow of the American Institute of Aeronautics and Astronautics; member of the Society of Experimental Test Pilots, Sigma Gamma Tau (aeronautical engineering society), Tau Beta Pi (national engineering society), and Sigma Xi (national science research society); and a 32 Degree Mason advanced through the Commandery and Shrine.
SPECIAL HONORS: Awarded the Distinguished Flying Cross with one oak leaf cluster, the Air Medal with two oak leaf clusters, the Air Force Commendation Medal, the NASA Exceptional Service Medal and Air Force Command Pilot Astronaut Wings, the NASA Group Achievement Award for Rendezvous Operations Planning Team, an Honorary Life Membership in the International Association of Machinists and Aerospace Workers, and an Honory Membership in the Aerospace Medical Association.
EXPERIENCE: Aldrin, an Air Force Colonel, was graduated third in a class of 475 from the United States Military Academy at West Point in 1951 and subsequently received his wings at Bryan, Texas, in 1952.

He flew 66 combat missions in F-86 aircraft while on duty in Korea with the 51st Fighter Interceptor Wing and was credited with destroying two MIG-15 aircraft. At Nellis Air Force Base, Nevada, he served as an aerial gunnery instructor and then attended the Squadron officers' School at the Air University, Maxwell Air Force base, Alabama.

Following his assignment as Aide to the Dean of Faculty at the United States Air Force Academy, Aldrin flew F-100 aircraft as a flight commander with the 36th Tactical Fighter Wing at Bitburg, Germany. He attended MIT, receiving a doctorate after completing his thesis concerning guidance for manned orbital rendezvous, and was then assigned to the Gemini Target Office of the Air Force Space Systems Division, Los Angeles, California. He was later transferred to the USAF Field Office at the Manned Spacecraft Center which was responsible for integrating DOD experiments into the NASA Gemini flights.

He has logged approximately 3500 hours flying time, including 2853 hours in jet aircraft and 139 hours in helicopters. He has made several flights in the lunar landing research vehicle.

CURRENT ASSIGNMENT: Colonel Aldrin was one of the third group of astronauts named by NASA in October 1963. He has since served as backup pilot for the Gemini 9 mission.

On 11 November 1966, he and command pilot James Lovell were launched in the Gemini 12 spacecraft on a 4-day, 59-revolution flight which brought the Gemini Program to a successful close. Aldrin established a new record for extravehicular activity (EVA) by accruing slightly more than 5½ hours outside the spacecraft.

Lunar Module Pilot (LMP)
NAME: Fred Wallace Haise, Jr. (Mr.)
DATE OF BIRTH: 14 November 1933
PHYSICAL DESCRIPTION: Height: 5 feet, 9-1/2 inches; weight: 150 pounds.
EDUCATION: Graduated from Biloxi High School, Biloxi, Mississippi; attended Perkinston Junior College (Association of Arts); received a Bachelor of Science degree with honors in Aeronautical Engineering from the University of Oklahoma in 1959.
ORGANIZATIONS: Member of the Society of Experimental Test Pilots, Tau Beta Pi, Sigma Gamma Tau, and Phi Theta Kappa.
SPECIAL HONORS: Recipient of the A.B. Honts Trophy as the outstanding graduate of class 64A from the Aerospace Research Pilot School in 1964; awarded the American Defense Ribbon and the Society of Experimental Test Pilots Ray E. Tenhoff Award for 1966.

BACKUP CREW

NEIL A. ARMSTRONG

EDWIN E. ALDRIN, Jr.

FRED W. HAISE, Jr.

EXPERIENCE: Haise was a research pilot at the NASA Flight Research Center at Edwards, California, before coming to Houston and the Manned Spacecraft Center; and from September 1959 to March 1963, he was a research pilot at the NASA Lewis Research Center in Cleveland, Ohio. During this time he authored the following papers which have been published: NASA TND, entitled "An Evaluation of the Flying Qualities of Seven General-Aviation Aircraft;" NASA TND 3380, "Use of Aircraft for Zero Gravity Environment, May 1966," SAE Business Aircraft Conference Paper, entitled "An Evaluation of General-Aviation A/C Flying Qualities,"

30 March - 1 April 1966; and a paper delivered at the tenth symposium of the Society of Experimental Test Pilots, entitled "A Quantitative Qualitative Handling Qualities Evaluation of Seven General-Aviation Aircraft," 1966.

He was the Aerospace Research Pilot School's outstanding graduate of class 64A and served with the US Air Force from October 1961 to August 1962 as a tactical fighter pilot and as Chief of the 164th Standardization-Evaluation Flight of the 164th Tactical Fighter Squadron at Mansfield, Ohio. From March 1957 to September

1959, he was a fighter-interceptor pilot with the 185th Fighter interceptor Squadron in the Oklahoma Air Notional Guard.

He also served as a tactics and all weather flight instructor in the US Navy Advanced Training Command at NAAS Kingsville, Texas, and was assigned as a US Marine Corps fighter pilot to VMF-533 and 114 at MCAS Cherry Point, North Carolina, from March 1954 to September 1956.

His military career began in October 1952 as a Naval Aviation Cadet at the Naval Air Station in Pensacola, Florida.

He has accumulated 5800 hours flying time, including 3000 hours in jets.

CURRENT ASSIGNMENT: Mr. Haise is one of the 19 astronauts selected by NASA in April 1966.

MISSION MANAGEMENT RESPONSIBILITY

Title	Name	Organization
Director, Apollo Program	Lt. Gen. Sam C. Phillips	NASA/OMSF
Director, Mission Operations	Maj. Gen. John D. Stevenson (Ret)	NASA/OMSF
Saturn V Vehicle Prog. Mgr.	Mr. Lee B. James	NASA/MSFC
Apollo Spacecraft Prog. Mgr.	Mr. George M. Low	NASA/MSC
Apollo Prog. Manager KSC	R. Adm. Roderick O. Middleton	NASA/KSC
Mission Director	Mr. William C. Schneider	NASA/OMSF
Assistant Mission Director	Capt. Chester M. Lee (Ret)	NASA/OMSF
Assistant Mission Director	Col. Thomas H. McMullen	NASA/OMSF
Director of Launch Operations	Mr. Rocco Petrone	NASA/KSC
Director of Flight Operations	Mr. Christopher C. Kraft	NASA/MSC
Launch Operations Manager	Mr. Paul C. Donnelly	NASA/KSC
Flight Directors	Mr. Clifford E. Charlesworth	NASA/MSC
	Mr. Glynn Lunney	
	Mr. Milton L. Windier	
Spacecraft Commander (Prime)	Col. Frank Borman	NASA/MSC
Spacecraft Commander (Backup)	Mr. Neil A. Armstrong	NASA/MSC

PROGRAM MANAGEMENT

NASA HEADQUARTERS
Office of Manned Space Flight
Manned Spacecraft Center
Marshall Space Flight Center
Kennedy Space Center

LAUNCH VEHICLE	SPACECRAFT	TRACKING AND DATA ACQUISITION
Marshall Space Flight Center The Boeing Co. (S-IC)	Manned Spacecraft Center	Kennedy Space Center
North American Rockwell (S-II)	North American Rockwell (LES, CSM, SLA)	Goddard Space Flight Center
McDonnell Douglas Corp. (S-IVB)	Manned Spacecraft Center (LTA)	Department of Defense
IBM Corp. (IU)		MSFN

ABBREVIATIONS

AC	Alternating Current
AK	Apogee Kick
APS	Ascent Propulsion System
BMAG	Body Mounted Attitude Gyro
CCATS	Communications, Command and Telemetry System
CCS	Command Communication System
CDR	Command Pilot
CM	Command Module
CMC	Command Module Computer
CMP	Command Module Pilot
COI	Contingency Orbit Insertion
CSM	Command and Service Module
C/T	Crawler Transporter
DOD	Department of Defense
DPS	Descent Propulsion System
DSIF	Deep Space Instrumentation Facility
DSKY	Display and Keyboard
ECS	Environmental Control System
EDS	Emergency Detection System
EI	Entry Interface
ELS	Earth Landing System
EPS	Electrical Power System
EST	Eastern Standard Time
FDAI	Flight Director Attitude Indicator
G&N	Guidance and Navigation
GDS	Goldstone
GET	Ground Elapsed Time
GNCS	Guidance, Navigation and Control System
H2	Hydrogen
H2O	Water
HSK	Honeysuckle Creek
IMU	Inertial Measurement Unit
ISS	Inertial Subsystem
IU	Instrument Unit
KSC	Kennedy Space Center
LC	Launch Complex
LCC	Launch Control Center
LEA	Launch Escape Assembly
LES	Launch Escape System
LET	Launch Escape Tower
LM	Lunar Module
LMP	Lunar Module Pilot
LOI	Lunar Orbit Insertion
LOR	Lunar Orbit Rendezvous
LOX	Liquid Oxygen
LTA	Lunar Module Test Article
LV	Launch Vehicle
MAD	Madrid
MCC	Midcourse Correction
MCC	Mission Control Center
ML	Mobile Launcher
MOCR	Mission Operation Control Room
MOR	Mission Operation Report
MORS	Mission Operation Report Supplement

MSFN	Manned Space Flight Network
MSS	Mobile Service Structure
NASCOM	NASA Communications
NAV	Navigation
O2	Oxygen
OSS	Operational Support System
PRS	Primary Recovery Ship
PTC	Passive Thermal Control
PTP	Preferred Target Point
RCS	Reaction Control System
RHC	Rotational Hand Controller
RSO	Range Safety Officer
RTCC	Real Time Computer Complex
SCS	Stabilization and Control System
SECS	Sequential Events Control System
SEQ	Sequential
SLA	Spacecraft LM Adapter
SM	Service Module
SMJC	Service Module Jettison Controller
SPS	Service Propulsion System
SV	Space Vehicle
T/C	Telecommunications
TEC	Transearth Coast
TEI	Transearth Injection
TLC	Translunar Coast
TLI	Translunar Injection
TVC	Thrust Vector Control
VAB	Vehicle Assembly Building

Mission Operation Report Supplement

Apollo 8 (AS.503) Mission

19 DECEMBER 1968

FOREWORD

MISSION OPERATION REPORTS are published expressly for the use of NASA Senior Management, as required by the Administrator in NASA Instruction 6-2-10, dated August 15, 1963. The purpose of these reports is to provide NASA Senior Management with timely, complete, and definitive information on flight mission plans, and to establish official mission objectives which provide the basis for assessment of mission accomplishment.

Initial reports are prepared and issued for each flight project just prior to launch. Following launch, updating reports for each mission are issued to keep General Management currently informed of definitive mission results as provided in NASA Instruction 6-2-10. Because of their sometimes highly technical orientation, distribution of these reports is limited to personnel having program-project management responsibilities. The Office of Public Affairs publishes a comprehensive series of pre-launch and post-launch reports on NASA flight missions, which are available for general distribution.

THE MISSION OPERATION REPORT SUPPLEMENT is intended to discuss facilities and equipments common to all Apollo-Saturn V missions. This supplement allows an adequate description of facilities and equipments without detracting from the unique mission information contained in each basic Mission Operation Report.

Specific technical data has been referenced to the SA-504 launch vehicle and the Block II spacecraft as a baseline.

SPACE VEHICLE

The primary flight hardware of the Apollo program consists of a Launch Vehicle (LV), designated the Saturn V, and an Apollo Spacecraft. Collectively, they are designated the Space Vehicle (SV) (Figure 1).

The Saturn Launch Vehicle consists of three propulsive stages (S-IC, S-II, S-IVB), two interstages, and an Instrument Unit (IU).

The Apollo Spacecraft (SC) includes the Spacecraft/Lunar Module Adapter (SLA), the Lunar Module (LM), the Service Module (SM), the Command Module (CM), and the Launch Escape System (LES). The Command Module and the Service Module, considered as a unit, is termed the Command/Service Module (CSM).

S-IC STAGE GENERAL

The S-IC stage (Figure 2) is a large cylindrical booster, 138 feet long and 33 feet in diameter, powered by five liquid propellant F-1 rocket engines. These engines develop a nominal sea level thrust of 1,526,500 pounds each (approximately 7,632,500 pounds total) and have a burn time of 150.5 seconds. The stage dry weight is approximately 295,300 pounds and the total loaded stage weight is approximately 5,030,300 pounds.

The stage interfaces structurally and electrically with the S-II stage. It also interfaces structurally, electrically, and pneumatically with Ground Support Equipment (GSE) including two umbilical service arms, three tail service masts, and certain electronic systems by antennas.

The stage consists of structures, propulsion, environmental control, fluid power, pneumatic control, propellants, electrical, instrumentation, and ordnance systems.

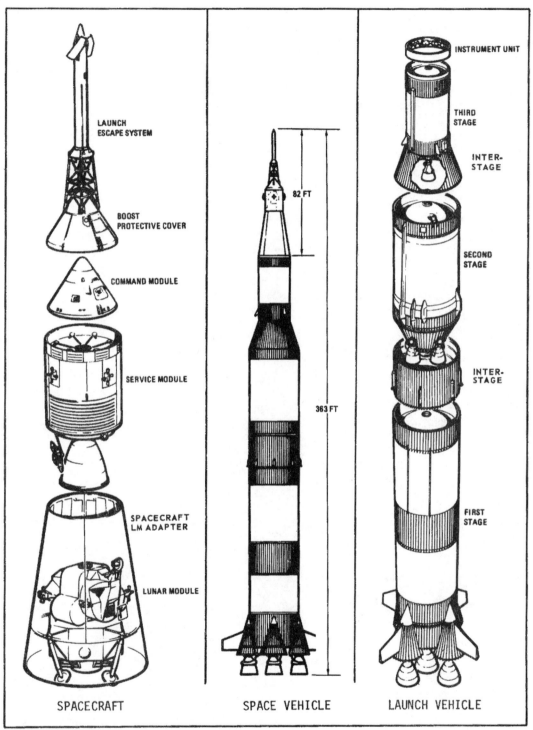

LAUNCH
ESCAPE SYSTEM

BOOST
PROTECTIVE COVER

COMMAND MODULE

SERVICE MODULE

SPACECRAFT
LM ADAPTER

LUNAR MODULE

82 FT

363 FT

INSTRUMENT UNIT

THIRD
STAGE

INTER-
STAGE

SECOND
STAGE

INTER-
STAGE

FIRST
STAGE

SPACECRAFT   SPACE VEHICLE   LAUNCH VEHICLE

Fig. 1

STRUCTURE

The S-IC structural design reflects the requirements of F-I engines, propellants, control, instrumentation, and interfacing systems. Aluminum alloy is the primary structural material. The major components, shown in Figure 3 are the forward skirt, oxidizer tank, intertank section, fuel tank, and thrust structure.

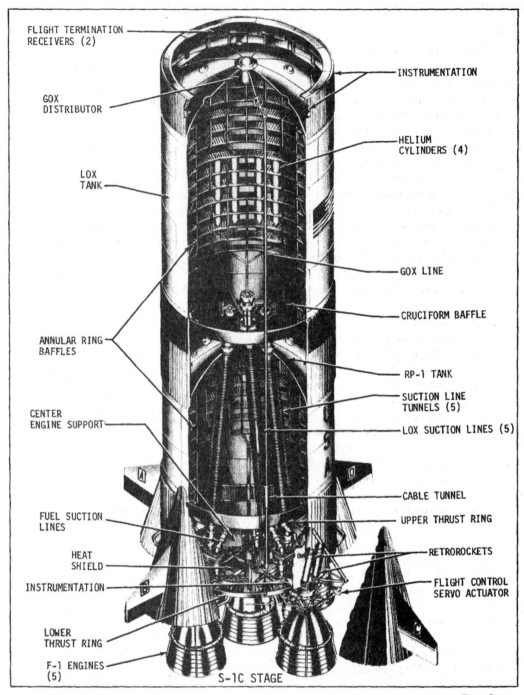

FLIGHT TERMINATION
RECEIVERS (2)

GOX
DISTRIBUTOR

LOX
TANK

ANNULAR RING
BAFFLES

CENTER
ENGINE SUPPORT

FUEL SUCTION
LINES

HEAT
SHIELD

INSTRUMENTATION

LOWER
THRUST RING

F-1 ENGINES
(5)

INSTRUMENTATION

HELIUM
CYLINDERS (4)

GOX LINE

CRUCIFORM BAFFLE

RP-1 TANK

SUCTION LINE
TUNNELS (5)

LOX SUCTION LINES (5)

CABLE TUNNEL

UPPER THRUST RING

RETROROCKETS

FLIGHT CONTROL
SERVO ACTUATOR

S-1C STAGE

Fig. 2

The 345,000-gallon oxidizer tank is the structural link between the forward skirt and the intertank structure which provides structural continuity between the oxidizer and fuel tanks.

The 216,000-gallon fuel tank (Figure 3) provides the load carrying structural link between the thrust and intertank structures. It is cylindrical with ellipsoidal upper and lower bulkheads. Five oxidizer ducts run from the oxidizer tank, through the fuel tank, to the F-1 engines.

The thrust structure assembly (Figure 3) redistributes the applied loads of the five F-1 engines into nearly uniform loading about the periphery of the fuel tank. Also, it provides support for the five F-1 engines, engine accessories, base heat shield engine fairings and fins, propellant lines, retrorockets, and environmental control ducts. The lower thrust ring has four holddown points which support the fully loaded Saturn V/ Apollo (approximately 6,000,000 pounds) and also, as necessary, restrain the vehicle during controlled release.

PROPULSION

The F-1 engine is a single-start 1,526,500 pound fixed-thrust, calibrated, bi-propellant engine which uses liquid oxygen (LOX) as the oxidizer and Rocket Propellant-1 (RP-1) as the fuel. The thrust chamber is cooled regeneratively by fuel, and the nozzle extension is cooled by gas generator exhaust gases LOX and RP-1 are supplied to the thrust chamber by a single turbopump powered by a gas generator which uses the some propellant combination RP-1 is also used as the turbopump lubricant and as the working fluid for the engine hydraulic control system. The four outboard engines are capable of gimbaling and have provisions for supply and return of RP-1 as the working fluid for a thrust vector control system. The engine contains a heat exchanger system to condition engine supplied LOX and externally supplied helium for stage propellant tank pressurization. An instrumentation system monitors engine performance and operation. External thermal insulation provides an allowable engine environment during flight operation.

The nominal inflight engine cutoff sequence is center engine first, followed by the four outboard engines. Engine optical-type LOX depletion sensors initiate the engine cutoff sequence. A fuel level engine cutoff sensor in the bottom of the fuel tank initiates engine shutdown when RP-1 is depleted if the LOX sensors have failed to cut the engines off first.

In an emergency, the engine can be cut off by any of the following methods: Ground Support Equipment (GSE) Command Cutoff, Emergency Detection System, Outboard Cutoff System.

PROPELLANT SYSTEMS

The systems include hardware for fill and drain, propellant conditioning, and tank pressurization prior to and during flight, and for delivery to the engines.

Fuel tank pressurization is required during engine starting and flight to establish and maintain a net positive suction head (NPSH) at the fuel inlet to the engine turbo pumps. During flight, the source of fuel tank pressurization is helium from storage bottles mounted inside the oxidizer tank.

Fuel feed is accomplished through two 12-inch ducts which connect the fuel tank to each F-1 engine. The ducts are equipped with flex and sliding joints to compensate for motions from engine gimbaling and stage stresses.

Gaseous oxygen (GOX) is used for oxidizer tank pressurization during flight. A portion of the LOX supplied to each engine is diverted into the engine heat exchangers where it is transformed into GOX and routed back to the tanks.

LOX is delivered to the engines through five suction lines which are equipped with flex and sliding joints.

FLIGHT CONTROL SYSTEM

The S-IC thrust vector control consists of four outboard F-1 engines, gimbal blocks to attach these engines to the thrust ring, engine hydraulic servo-actuators (two per engine), and an engine hydraulic power supply.

Engine thrust is transmitted to the thrust structure through the engine gimbal block. There are two servo-actuator attach points per engine, located 90 degrees from each other, through which the gimbaling force is applied. The gimbaling of the four outboard engines changes the direction of thrust and as a result corrects the attitude of the vehicle to achieve the desired trajectory.

ELECTRICAL

The electrical power system of the S-IC stage is made up of two basic subsystems: the operational power subsystem and the measurements power subsystem. Onboard power is supplied by two 28-volt batteries. Battery number 1 is identified as the operational power system battery. It supplies power to operational loads such as valve controls, purge and venting systems, pressurization systems, and sequencing and flight control. Battery number 2 is identified as the measurement power system. Batteries supply power to their loads through a common main power distributor, but each system is completely isolated from the other. The S-IC stage switch selector is the interface between the Launch Vehicle Digital Computer (LVDC) in the IU and the S-IC stage electrical circuits. Its function is to sequence and control various flight activities such as telemetry calibration, retrofire initiation, and pressurization.

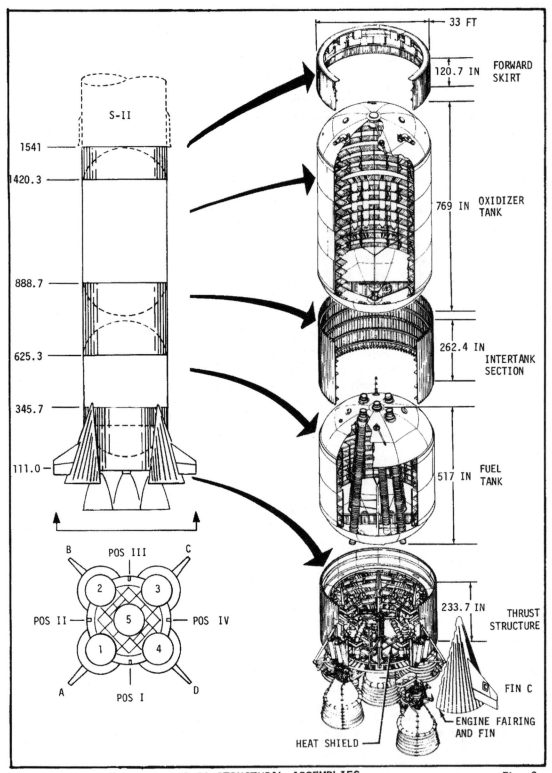

S-IC STRUCTURAL ASSEMBLIES   Fig. 3

ORDNANCE

The S-IC ordnance systems include the propellant dispersion (flight termination) system and the retrorocket system.

The S-IC propellant dispersion system(PDS) provides the means of terminating the flight of the Saturn V if it varies beyond the prescribed limits of its flight path or if it becomes a safety hazard during the

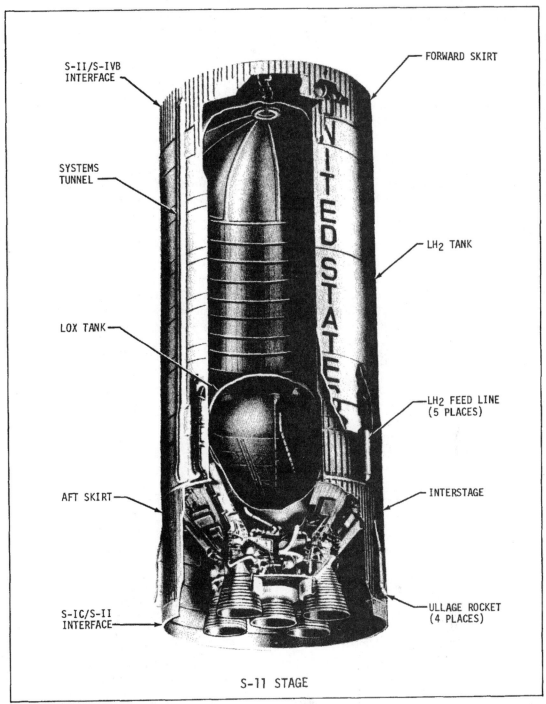

FORWARD SKIRT

S-II/S-IVB INTERFACE

SYSTEMS TUNNEL

LH₂ TANK

LOX TANK

LH₂ FEED LINE (5 PLACES)

AFT SKIRT

INTERSTAGE

S-IC/S-II INTERFACE

ULLAGE ROCKET (4 PLACES)

S-11 STAGE

Fig. 4

S-IC boost phase.

The eight retrorockets that provide separation thrust after S-IC burnout are attached externally to the thrust structure inside the four outboard engine fairings.

The S-IC retrorockets are mounted, in pairs, in the fairings of the F-1 engines. The firing command originates in the IU and activates redundant firing systems. At retrorocket ignition the forward end of the fairing is burned and blown through by the exhausting gases. The thrust level developed by seven retrorockets (one retrorocket out) is adequate to separate the S-IC stage a minimum of six feet from the vehicle in less than one second.

S-II STAGE GENERAL

The S-II stage (Figure 4) is a large cylindrical booster, 81.5 feet long and 33 feet in diameter, powered by five liquid propellant J-2 rocket engines which develop a nominal vacuum thrust of 230,000 pounds each for a total of 1,150,000 pounds.

Dry weight of the S-II stage is approximately 85,522 pounds (95,779 pounds including the S-IC/S-II interstage). The stage approximate loaded gross weight is 1,080,500 pounds. The four outer J-2 engines are equally spaced on a 17.5-foot diameter circle and are capable of being gimbaled through a plus or minus 7.0 degree square pattern for thrust vector control. The fifth engine is mounted on the stage centerline and is fixed.

The stage (Figure 5) consists of the following major systems: structural, environmental control, propulsion, flight control, pneumatic, propellant, electrical, instrumentation and ordnance. The stage has structural and electrical interfaces with the S-IC and S-IVB stages; and electric, pneumatic, and fluid interfaces with GSE through its umbilicals and antennas.

STRUCTURE

The S-II airframe (Figure 5) consists of a body shell structure (forward and aft skirts and interstage), a propellant tank structure (fuel and oxidizer tanks), and a thrust structure. The body shell structure transmits boost loads and stage body bending and longitudinal forces between the adjacent stages, the propellant tank structure, and the thrust structure. The propellant tank structure holds the propellants, liquid hydrogen (LH2) and liquid oxygen (LOX), and provides structural support between the aft and forward skirts. The thrust structure transmits the thrust of the five J-2 engines to the body shell structure; compression loads from engine thrust; tension loads from idle engine weight; and cantilever loads from engine weight during S-II boost.

PROPULSION

The S-II stage engine system consists of five single-start, high performance, high altitude J-2 rocket engines. The four outer J-2 engines are suspended by gimbal bearings to allow thrust vector control. The fifth engine is fixed and is mounted on the centerline of the stage.

The engine valves are controlled by a pneumatic system powered by gaseous helium which is stored in a sphere inside the start tank. An electrical control system, which uses solid state logic elements, is used to sequence the start and shutdown operations of the engine. Electrical power is stage supplied.

The J-2 engine may receive cutoff signals from several different sources. These sources include engine interlock deviations, EDS automatic manual abort cutoffs and propellant depletion cutoff. Each of these sources signal the LVDC in the IU. The LVDC sends the engine cutoff signal to the S-II switch selector, which in turn signals the electrical control package, which signals for the cutoff sequence.

Five discrete liquid level sensors per propellant tank provide initiation of engine cutoff upon detection of propellant depletion. The cutoff sensors will initiate a signal to shut down the engines when two out of five engine cutoff signals from the same tank are received.

PROPELLANT SYSTEMS

The propellant systems supply fuel and oxidizer to the five engines. This is accomplished by the propellant management components and the servicing, conditioning, and engine delivery subsystems.

The LH2 feed system includes five 8-inch vacuum-jacketed feed ducts and five prevalves.

During powered flight, prior to S-II ignition, gaseous hydrogen (GH2) for LH2 tank pressurization is bled from the thrust chamber hydrogen injector manifold of each of the four outboard engines. After S-II engine ignition, LH2 is preheated in the regenerative cooling tubes of the engine and tapped off from the thrust chamber injector manifold in the form of GH2 to serve as a pressurizing medium.

The LOX feed system includes four 8-inch vacuum jacketed feed ducts, one uninsulated feed duct and five prevalves.

LOX tank pressurization is accomplished with GOX obtained by heating LOX bled from the LOX turbopump outlet.

The propellant management system monitors propellant mass for control of propellant loading, utilization, and depletion. Components of the system include continuous capacitance probes, propellant utilization valves, liquid level sensors, and electronic equipment.

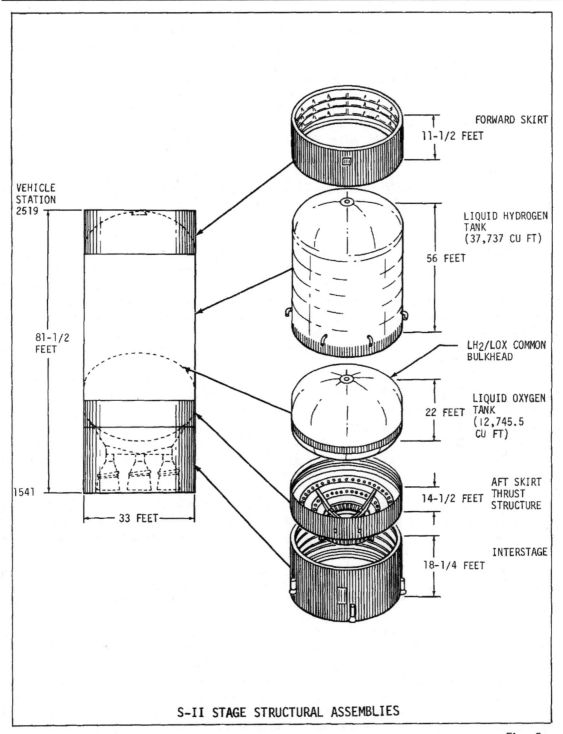

S-II STAGE STRUCTURAL ASSEMBLIES

Fig. 5

During flight, the signals from the tank continuous capacitance probes are monitored and compared to provide an error signal to the propellant utilization valve an each LOX pump. Based on this error signal, the propellant utilization valves are positioned to minimize residual propellants and assure a fuel-rich cutoff by varying the amount of LOX delivered to the engines.

FLIGHT CONTROL SYSTEM

Each outboard engine is equipped with a separate, independent, closed-loop, hydraulic control system that includes two servo-actuators mounted in perpendicular planes to provide control over the vehicle pitch, roll and yaw axes. The servo-actuators are capable of deflecting the engine $\pm$ 7 degrees in the pitch and yaw planes, + 10 degrees diagonally, at the rate of 8 degrees per second.

ELECTRICAL

The electrical system is comprised of the electrical power and electrical control subsystems. The electrical power system provides the S-II stage with the electrical power source and distribution. The electrical control system interfaces with the IU to accomplish the mission requirements of the stage. The LVDC in the IU controls inflight sequencing of stage functions through the stage switch selector. The stage switch selector outputs are routed through the stage electrical sequence controller or the separation controller to accomplish the directed operation. These units are basically a network of low-power, transistorized switches that can be controlled individually and, upon command from the switch selector, provide properly sequenced electrical signals to control the stage functions.

ORDNANCE

The S-II ordnance systems include the separation, usage rocket, retrorocket, and propellant dispersion (flight termination) systems.

For S-IC/S-II separation, a dual plane separation technique is used wherein the structure between the two stages is severed at two different planes. The S-II/S-IVB separation occurs at a single plane. All separations are controlled by the LVDC located in the IU.

To ensure stable flow of propellants into the J-2 engines, a small forward acceleration is required to settle the propellants in their tanks. This acceleration is provided by four ullage rockets.

To separate and retard the S-II stage, a deceleration is provided by the four retrorockets located in the S-II/S-IVB interstage.

The S-II PDS provides for termination of vehicle flight during the S-II boost phase if the vehicle flight path varies beyond its prescribed limits or if continuation of vehicle flight creates a safety hazard. The S-II PDS may be safed after the launch escape tower is jettisoned.

The LH2 tank linear shaped charge, when detonated, cuts a 30-foot vertical opening in the tank.

The LOX tank destruct charges simultaneously cut 13-foot lateral openings in the LOX tank and the S-II aft skirt.

S-IVB STAGE GENERAL

The S-IVB (Figure 6) is the third launch vehicle stage. Its single J-2 engine is designed to boost the payload into a circular earth orbit on the first burn then inject the payload into the trajectory for lunar intercept with a second burn. The subsystems to accomplish this mission are described in the following section.

STRUCTURE

The S-IVB stage is a bi-propellant tank structure, designed to withstand the normal loads and stresses incurred on the ground and during launch, pre-ignition boost, ignition, and all other flight phases.

The basic S-IVB stage airframe, illustrated in Figure 7, consists of the following structural assemblies: the forward skirt, propellant tanks, aft skirt, thrust structure, and aft interstage. These assemblies, with the exception of the propellant tanks, are all of a skin/stringer-type aluminum alloy airframe construction. In addition, there are two longitudinal tunnels which house wiring, pressurization lines, and propellant dispersion systems.

The cylindrical forward skirt extends forward from the intersection of the fuel tank sidewall and the forward dome, serving as a load-supporting member between the fuel tank and the IU.

The propellant tank assembly consists of a cylindrical tank with a hemispherical shaped dome at each end, and a common bulkhead to separate the fuel (LH2) from the oxidizer (LOX). This bulkhead is of sandwich-type construction, consisting of two parallel hemispherical shaped aluminum alloy domes bonded to and separated by a fiberglass phenolic honeycomb core.

Attached to the inside of the fuel tank are a 34-foot propellant utilization (PU) probe, nine cold helium spheres, brackets with temperature and level sensors, a chilldown pump, a slosh baffle, a slosh deflector, and fill, pressurization, and vent pipes. Attached to the inside of the oxidizer tank are slosh baffles, a chilldown pump, a 13.5-foot PU probe, temperature and level sensors, and fill, pressurization and vent pipes.

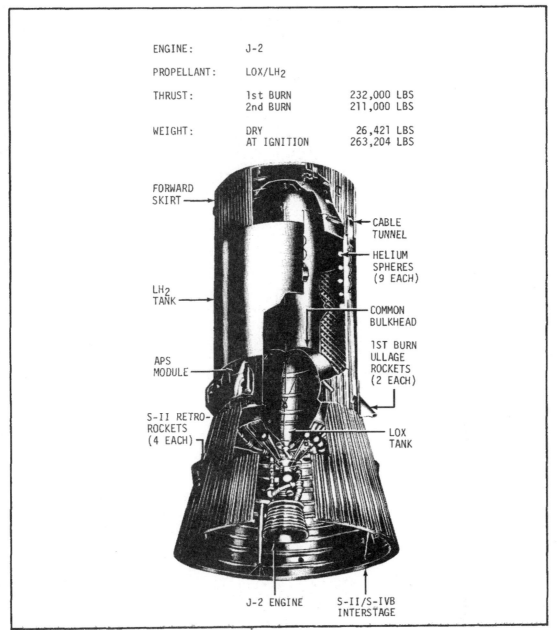

```
ENGINE:          J-2

PROPELLANT:      LOX/LH2

THRUST:          1st BURN          232,000 LBS
                 2nd BURN          211,000 LBS

WEIGHT:          DRY                26,421 LBS
                 AT IGNITION       263,204 LBS
```

FORWARD
SKIRT

LH2
TANK

APS
MODULE

S-II RETRO-
ROCKETS
(4 EACH)

CABLE
TUNNEL

HELIUM
SPHERES
(9 EACH)

COMMON
BULKHEAD

1ST BURN
ULLAGE
ROCKETS
(2 EACH)

LOX
TANK

J-2 ENGINE   S-II/S-IVB
                    INTERSTAGE

S-1VB STAGE   Fig. 6

The thrust structure assembly is an inverted, truncated cone attached at its large end to the aft dome of the oxidizer tank and attached at its small end to the engine mount. It provides the attach point for the J-2 engine.

The aft skirt assembly is the load-bearing structure between the fuel tank and aft interstage. It is bolted to the tank assembly at its forward edge and connected to the aft interstage with a frangible tension tie which separates the S-IVB from the aft interstage.

The aft interstage is a truncated cone that provides the load supporting structure between the S-IVB stage and the S-II stage. The interstage also provides the focal point for the required electrical and mechanical interface between the S-II and S-IVB stages. The S-II retrorocket motors are attached to this interstage and at separation the interstage remains attached to the S-II.

MAIN PROPULSION

The high performance J-2 engine as installed in the S-IVB stage has a multiple restart capability. The engine valves are controlled by a pneumatic system powered by gaseous helium which is stored in a sphere

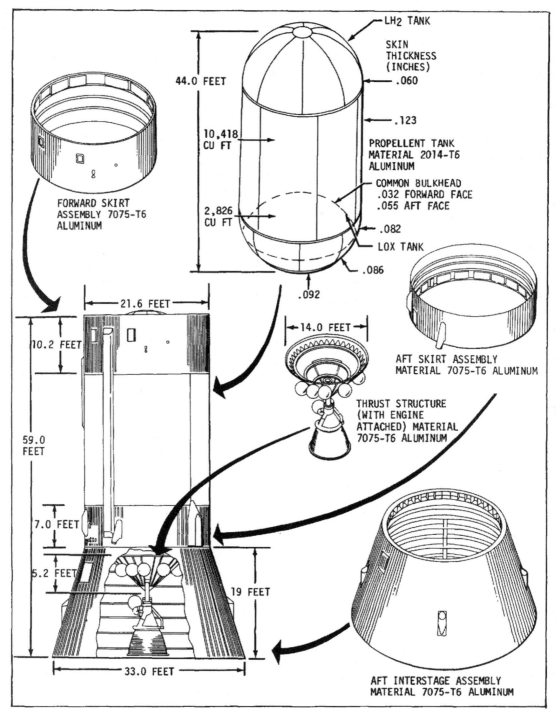

S-IVB STRUCTURAL ASSEMBLIES   Fig. 7

inside a start bottle. An electrical control system, which uses solid state logic elements, is used to sequence the start and shutdown operations of the engine. Electrical power is supplied from aft battery No. 1.

During the burn periods, the oxidizer tank is pressurized by flowing cold helium through the heat exchanger in the oxidizer turbine exhaust duct. The heat exchanger heats the cold helium, causing it to expand. The fuel tank is pressurized during burn periods by GH2 from the thrust chamber fuel manifold.

Thrust vector control, in the pitch and yaw planes, during burn periods is achieved by gimbaling the entire engine.

The J-2 engine may receive cutoff signals from the following sources; EDS, range safety systems,

"Thrust OK" pressure switches, propellant depletion sensors, and an IU programmed command (velocity or timed) via the switch selector.

The restart of the J-2 engine is identical to the initial start except for the fill procedure of the start tank. The start tank is filled with LH2 and GH2 during the first burn period by bleeding GH2 from the thrust chamber fuel injection manifold and LH2 From the Augmented Spark Igniter (ASI) fuel line to refill the start tank for engine restart. (Approximately 50 seconds of mainstage engine operation is required to recharge the start tank.)

To insure that sufficient energy will be available for spinning the LH2 and LOX pump turbines, a waiting period of between approximately 90 minutes to 6 hours is required. The minimum time is required to build sufficient pressure by warming the start tank through natural means and to allow the hot gas turbine exhaust system to cool. Prolonged heating will cause a loss of energy in the start tank. This loss occurs when the LH2 and GH2 warms and raises the gas pressure to the relief valve setting. If this venting continues over a prolonged period the total stored energy will be depleted. This limits the waiting period prior to a restart attempt to six hours.

PROPELLANT SYSTEMS

LOX is stored in the aft tank of the propellant tank structure at a temperature of -297°F.

A six-inch, low-pressure supply duct supplies LOX from the tank to the engine. During engine burn, LOX is supplied at a nominal flow rate of 392 pounds per second, and at a transfer pressure above 25 psia. The supply duct is equipped with bellows to provide compensating flexibility for engine gimbaling, manufacturing tolerances, and thermal movement of structural connection.

The tank is prepressurized to between 38 and 41 psia and is maintained at that pressure during boost and engine operation. Gaseous helium is used as the pressurizing agent.

The LH2 is stored in an insulated tank at less than -423° F. LH2 from the tank is supplied to the J-2 engine turbopump by a vacuum-jacketed, low-pressure, 10-inch duct. This duct is capable of flowing 80-pounds per second at -423° F. and at a transfer pressure of 28 psia. The duct is located in the aft tank side wall above the common bulkhead joint. Bellows in this duct compensate for engine gimbaling, manufacturing tolerances, and thermal motion.

The fuel tank is prepressurized to 28 psia minimum and 31 psia maximum.

The PU subsystem provides a means of controlling the propellant mass ratio. It consists of oxidizer and fuel tank mass probes, a PU valve, and an electronic assembly. These components monitor the propellant and maintain command control.

Propellant utilization is provided by bypassing oxidizer from the oxidizer turbopump outlet back to the inlet. The PU valve is controlled by signals from the PU system. The engine oxidizer/fuel mixture mass ratio varies from 4.5:1 to 5.5:1.

FLIGHT CONTROL

The flight control system incorporates two systems for flight and attitude control. During powered flight, thrust vector steering is accomplished by gimbaling the J-2 engine for pitch and yaw control and by operating the Auxiliary Propulsion System (APS) engines for roll control. Steering during coast flight is by use of the APS engines alone.

The engine is gimbaled in a ± 7.5 degree square pattern by a closed-loop hydraulic system. Mechanical feedback from the actuator to the servovalve provides the closed engine position loop.

Two actuators are used to translate the steering signals into vector forces to position the engine. The deflection rates are proportional to the pitch and yaw steering signals from the flight control computer.

AUXILIARY PROPULSION SYSTEM

The S-IVB APS provides three-axis stage attitude control (Figure 8) and main stage propellant control during coast flight.

The APS engines are located in two modules 180° apart on the aft skirt of the S-IVB stage (Figure 9). Each module contains four engines; three 150-pound thrust control engines, and one 70-pound thrust ullage engine. Each module contains its own oxidizer, fuel, and pressurization system.

A positive expulsion propellant feed subsystem is used to assure that hypergolic propellants are supplied to the engines under "zero g" or random gravity conditions. Nitrogen tetroxide (N2O4), is the oxidizer and monomethyl hydrazine (MMH), is the fuel for these engines.

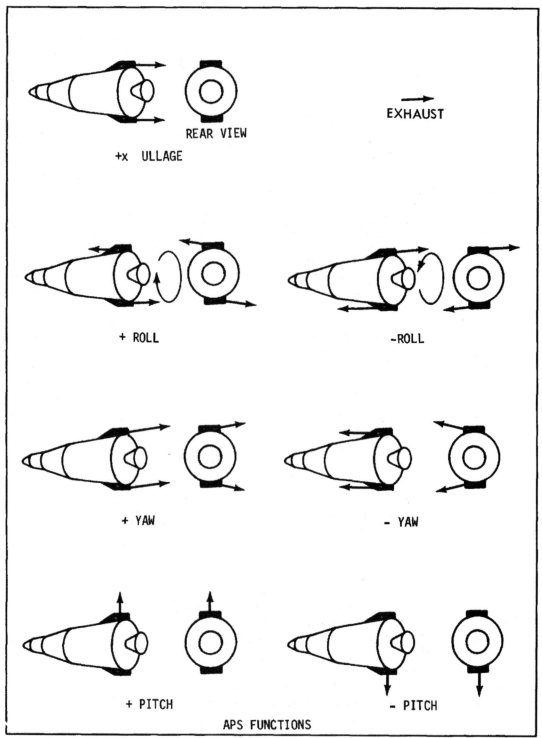

REAR VIEW

+x   ULLAGE

EXHAUST

+ ROLL

−ROLL

+ YAW

− YAW

+ PITCH

− PITCH

APS FUNCTIONS

Fig. 8

ELECTRICAL

The electrical system of the S-IVB stage is comprised of two major subsystems: the electrical power subsystem which consists of all the power sources on the stage; and the electrical control subsystem which distributes power and control signals to various loads throughout the stage.

Onboard electrical power is supplied by four silver-zinc batteries. Two are located in the forward equipment area and two in the aft equipment area. These batteries are activated and installed in the stage

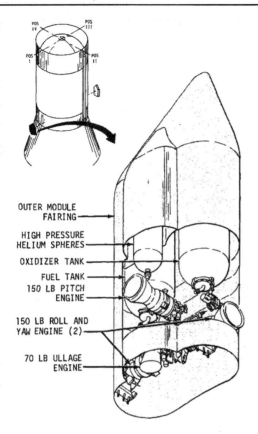

OUTER MODULE
FAIRING

HIGH PRESSURE
HELIUM SPHERES

OXIDIZER TANK

FUEL TANK
150 LB PITCH
ENGINE

150 LB ROLL AND
YAW ENGINE (2)

70 LB ULLAGE
ENGINE

AUXILIARY PROPULSION SYSTEM CONTROL MODULE

Fig. 9

during the final pre-launch preparations. Heaters and instrumentation probes are an integral part of each battery.

ORDNANCE

The S-IVB ordnance systems include the separation, ullage rocket, and PDS systems.

The separation system for S-II/S-IVB is located at the top of the S-II/S-IVB interstage.

At the time of separation, four retrorocket motors mounted on the interstage structure below the separation plane fire to decelerate the S-11 stage.

To provide propellant settling and thus ensure stable flow of fuel and oxidizer during J-2 engine start, the S-IVB stage requires a small acceleration. This acceleration is provided by two ullage rockets.

The S-IVB PDS provides for termination of vehicle flight. The S-IVB PDS may be safed after the launch escape tower is jettisoned.

Following S-IVB engine cutoff at orbit insertion, the PDS is electrically safed by ground command.

INSTRUMENT UNIT GENERAL

The Instrument Unit (IU) is a cylindrical structure 21.6 ft. in diameter and 3 ft. high installed on top of the S-IVB stage (Figure 10).The IU contains the guidance, navigation, and control equipment for the Launch Vehicle. In addition, it contains measurements and telemetry, command communications, tracking, and emergency detection system components along with supporting electrical power and environmental control systems.

STRUCTURE

The basic IU structure is a short cylinder fabricated of an aluminum alloy honeycomb sandwich material. Attached to the inner surface of the cylinder are "cold plates" which serve both as mounting structure and thermal conditioning units for the electrical/ electronic equipment.

NAVIGATION, GUIDANCE, AND CONTROL

The Saturn V Launch Vehicle is guided from its launch pad into earth orbit by navigation, guidance, and control equipment located in the IU. An all-inertial system using a space-stabilized platform for acceleration and attitude measurements is utilized. A Launch Vehicle Digital Computer (LVDC) is used to solve guidance equations and a Flight Control Computer (analog) is used for the flight control functions.

The three-gimbal stabilized platform (ST-124-M3) provides a space-fixed coordinate reference frame for attitude control and for navigation (acceleration) measurements. Three integrating accelerometers,

SATURN INSTRUMENT UNIT

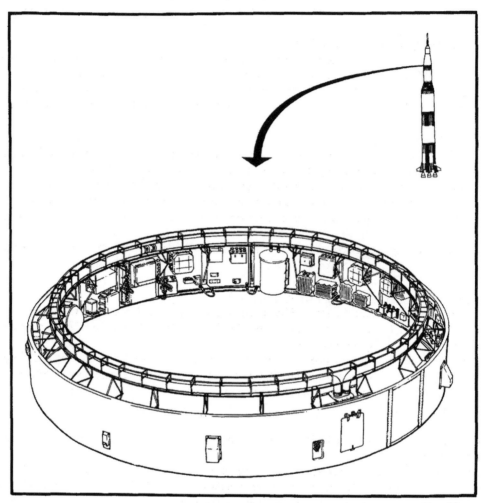

Fig. 10

mounted on the gyro-stabilized inner gimbal of the platform, measure the three components of velocity resulting from vehicle propulsion. The accelerometer measurements are sent through the Launch Vehicle Data Adapter (LVDA) to the LVDC. In the LVDC, the accelerometer measurements are combined with the computed gravitational acceleration to obtain velocity and position of the vehicle. During orbital flight, the navigational program continually computes the vehicle position, velocity, and acceleration. Guidance information stored in the LVDC (e.g., position, velocity) can be updated through the IU command system by data transmission from ground stations. The IU command system provides the general capability of changing or inserting information into the LVDC.

The control subsystem is designed to maintain and control vehicle attitude by forming the steering commands to be used by the controlling engines of the active stage.

The control system accepts guidance computations from the LVDC/LVDA Guidance System. These guidance commands, which are actually attitude error signals, are then combined with measured data from the various control sensors. The resultant output is the command signal to the various engine actuators and APS nozzles.

The final computations (analog) are performed within the flight control computer. This computer is also the central switching point for command signals. From this point, the signals are routed to their associated active stages and to the appropriate attitude control devices.

MEASUREMENTS AND TELEMETRY

The instrumentation within the IU consists of a measuring subsystem, a telemetry subsystem and an antenna subsystem. This instrumentation is for the purpose of monitoring certain conditions and events which take place within the IU and for transmitting monitored signals to ground receiving stations.

COMMAND COMMUNICATIONS SYSTEM

The Command Communications System (CCS) provides for digital data transmission from ground stations to the LVDC. This communications link is used to update guidance information or command certain other functions through the LVDC. Command data originates in the Mission Control Center and is sent to remote stations of the Manned Space Flight Network (MSFN) for transmission to the Launch Vehicle.

SATURN TRACKING INSTRUMENTATION

The Saturn V IU carries two C-band radar transponders and an Azusa/GLOTRAC tracking transponder. A combination of tracking data from different tracking systems provides the best possible trajectory information and increased reliability through redundant data. The tracking of the Saturn Launch Vehicle may be divided into 4 phases: powered flight into earth orbit, orbital flight, injection into mission trajectory, and coast flight after injection. Continuous tracking is required during powered flight into earth orbit.

During orbital flight, tracking is accomplished by S-band stations of the MSFN and by C-band radar stations.

IU EMERGENCY DETECTION SYSTEM COMPONENTS

The Emergency Detection System (EDS) is one element of several crew safety systems.

There are nine EDS rate gyros installed in the IU. Three gyros monitor each of the three axes, pitch, roll, and yaw, thus providing triple redundancy.

The control signal processor provides power to and receives inputs from the nine EDS rate gyros. These inputs are processed and sent on to the EDS distributor and to the flight control computer.

The EDS distributor serves as a junction box and switching device to furnish the spacecraft display panels with emergency signals if emergency conditions exist. It also contains relay and diode logic for the automatic abort sequence.

An electronic timer in the IU allows multiple engine shutdowns without automatic abort after 30 to 40 seconds of flight.

Inhibiting of automatic abort circuitry is also provided by the vehicle flight sequencing circuits through the IU switch selector. This inhibiting is required prior to normal S-IC engine cutoff and other normal vehicle sequencing. While the automatic abort is inhibited, the flight crew must initiate a manual abort if an angular-overrate or two engine-out condition arises.

ELECTRICAL POWER SYSTEMS

Primary flight power for the IU equipment is supplied by silver-zinc batteries at a nominal voltage level of 28 +2 vdc. Where ac power is required within the IU it is developed by solid state dc to ac inverters. Power distribution within the IU is accomplished through power distributors which are essentially junction boxes and switching circuits.

ENVIRONMENTAL CONTROL SYSTEM

The Environmental Control System (ECS) maintains an acceptable operating environment for the IU equipment during Preflight and flight operations. The ECS is composed of the following:

1. The Thermal Conditioning System (TCS) which maintains a circulating coolant temperature to the electronic equipment of $59° \pm 1°F$.

2. Pre-flight purging system which maintains a supply of temperature and pressure regulated air/gaseous nitrogen in the IU/S-IVB equipment area.

3. Gas bearing supply system which furnishes gaseous nitrogen to the ST-124-M3 inertial platform gas bearings.

4. Hazardous gas detection sampling equipment which monitors the IU/S-IVB forward interstage area for the presence of hazardous vapors.

SPACECRAFT LM ADAPTER GENERAL

The Spacecraft LM Adapter (SLA) is a conical structure which provides a structural load path between the LV and SM and also supports the LM. Aerodynamically, the SLA smoothly encloses the irregularly shaped LM and transitions the space vehicle diameter from that of the upper stage of the LV to that of the Service Module. The SLA also encloses the nozzle of the service module engine.

STRUCTURE

The SLA is constructed of 1.7-inch thick aluminum honeycomb panels. The four, upper, jettisonable or forward panels, are about 21 feet long, and the fixed lower or aft panels, about 7 feet long. The exterior surface of the SLA is covered completely by a layer of cork. The cork helps insulate the LM from aerodynamic heating during boost.

The LM is attached to the SLA at four locations around the lower panels.

SLA-SM SEPARATION

The SLA and SM are bolted together through flanges on each of the two structures. Explosive trains are used to separate the SLA and SM as well as for separating the four, upper, jettisonable SLA panels.

Redundancy is provided in three areas to assure separation; redundant initiating signals, redundant detonators and cord trains, and "sympathetic" detonation of nearby charges.

Pyrotechnic type and spring type thrusters are used in deploying and jettisoning the SLA upper panels. The four, double-piston pyrotechnic thrusters are located inside the SLA and start the panels swinging outward on their hinges. The two pistons of the thruster push on the ends of adjacent panels thus providing two separate thrusters operating each panel.

The explosive train which separates the panels is routed through two pressure cartridges in each thruster assembly.

The pyrotechnic thrusters route the panels 2 degrees establishing a constant angular velocity of 33 to 60 degrees per second. When the panels have rotated about 45 degrees, the partial hinges disengage and free the panels from the aft section of the SLA, subjecting them to the force of the spring thrusters. The spring thrusters are mounted on the outside of the upper panels. When the panel hinges disengage, the springs in the thruster push against the fixed lower panels to propel the panels away from the vehicle at an angle of 110 degrees to the centerline at a speed of about 5½ miles per hour. The panels will then depart the area of the spacecraft.

LAUNCH ESCAPE SYSTEM GENERAL

The Launch Escape System (LES) (Figure 11) includes the LES structure, canards, rocket motors, and ordnance. The LES provides an immediate means of separating the CM from the LV during pad or suborbital aborts up through completion of second stage ignition. During an abort, the LES must provide a satisfactory earth return trajectory and CM orientation before jettisoning from the CM. The jettison or abort can be initiated manually or automatically.

ASSEMBLY

The forward or rocket section of the system is cylindrical and houses three solid propellant rocket motors and a ballast compartment topped by a nose cone and "Q-ball", which measures attitude and flight dynamics of the space vehicle. The 500-pound tower is made of titanium tubes attached at the top to a structural skirt that covers the rocket exhaust nozzles and at the bottom to the CM by means of explosive bolts.

A Boost Protective Cover (BPC) is attached to the tower and completely covers the CM. It has 12 "blowout" ports for the CM reaction engines, vents, and an 8-inch window. This cover protects the CM from the rocket exhaust and also from the heat generated during launch vehicle boost. It remains attached to the tower and is carried away when the LES is jettisoned.

Two canards are deployed 11 seconds after an abort is initiated. The canards dynamically turn the CM so that the aft heat shield is forward. Three seconds later on extreme low-altitude aborts, or at approximately 24,000 feet on high-altitude aborts, the tower separation devices are fired and the jettison motor is started. These actions carry the LES away from the CM's landing trajectory. Four-tenths of a second after tower jettisoning the CM's earth landing system is activated and begins its sequence of operations to

LAUNCH ESCAPE SYSTEM

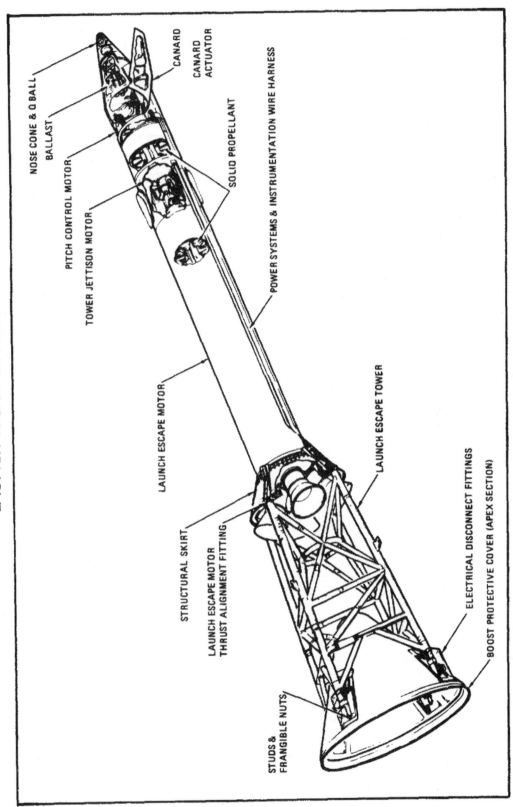

NOSE CONE & Q BALL

CANARD

BALLAST

CANARD ACTUATOR

PITCH CONTROL MOTOR

SOLID PROPELLANT

TOWER JETTISON MOTOR

POWER SYSTEMS & INSTRUMENTATION WIRE HARNESS

LAUNCH ESCAPE MOTOR

LAUNCH ESCAPE TOWER

STRUCTURAL SKIRT

LAUNCH ESCAPE MOTOR THRUST ALIGNMENT FITTING

ELECTRICAL DISCONNECT FITTINGS

BOOST PROTECTIVE COVER (APEX SECTION)

STUDS & FRANGIBLE NUTS

Fig. 11

bring the CM down safely.

During a successful launch the LES is jettisoned by the astronauts, using the digital events timer and the "S-II Sep" light as cues.

In the event of Tower Jettison Motor failure, the Launch Escape Motor may jettison the LES

PROPULSION

Three solid propellant motors are used on the LES. They are:

1. The Launch Escape Motor which provides thrust for CM abort. It weighs 4700 pounds and provides 147,000 pounds of thrust of sea level for approximately eight seconds.
2. The Pitch Control Motor which provides an initial pitch maneuver toward the Atlantic Ocean during pad or low-altitude abort. It weighs 50 pounds and provides 2400 pounds of thrust for a second.
3. The Tower Jettison Motor, which is used to jettison the LES, provides 31,500 pounds of thrust for one second.

SERVICE MODULE GENERAL

The Service Module (SM) (Figure 12) provides the main spacecraft propulsion and maneuvering capability during a mission. The Service Propulsion System (SPS) provides for velocity changes for translunar/ transearth course corrections, lunar orbit insertion, transearth injection and CSM aborts. The Service Module Reaction Control System (SM RCS) provides for maneuvering about and along three axes. The SM provides most of the spacecraft consumables (oxygen, water, propellant, hydrogen). It supplements environmental, electrical power and propulsion requirements of the CM. The SM remains attached to the CM until it is jettisoned just before CM entry.

STRUCTURE

The basic structural components are forward and aft (upper and lower) bulkheads, six radial beams, four sector honeycomb panels, four reaction control system honeycomb panels, and aft heat shield, and a fairing.

The radial beams are made of solid aluminum alloy which has been machined and chem-milled to thicknesses varying between 2 inches and 0.018 inch.

The forward and aft bulkheads cover the top and bottom of the SM. Radial beam trusses extending above the forward bulkhead support and secure the CM. Three of these beams have compression pads and the other three have shear-compression pads and tension ties. Explosive charges in the center sections of these tension ties are used to separate the CM from the SM.

An aft heat shield surrounds the service propulsion engine to protect the SM from the engine's heat during thrusting. The gap between the CM and the forward bulkhead of the SM is closed off with a fairing which is composed of eight electrical power system radiators alternated with eight aluminum honeycomb panels. The sector and reaction control system panels are 1-inch thick and are made of aluminum honeycomb core between two aluminum face sheets. The sector panels are bolted to the radial beams.

Radiators used to dissipate heat from the environmental control subsystem are bonded to the sector panels on opposite sides of the SM. These radiators are each about 30 square feet in area.

The SM interior is divided into six sectors and a center section (Figure 12b). Sector one is currently void. It is available for installation of scientific or additional equipment should the need arise.

Sector two has part of a space radiator and an RCS engine quad (module) on its exterior panel and contains the SPS oxidizer sump tank. This tank is the larger of the two tanks that hold the nitrogen tetroxide for the SPS engine.

Sector three has the rest of the space radiator and another RCS engine quad on its exterior panel and contains the oxidizer storage tank. This is the second of the two SPS oxidizer tanks and is fed from the oxidizer sump tank in sector two.

Sector four contains most of the electrical power generating equipment. It contains three fuel cells, two cryogenic oxygen and two cryogenic hydrogen tanks and a power control relay box. The cryogenic tanks supply oxygen to the environmental control subsystem and oxygen and hydrogen to the fuel cells.

Sector five has part of an environmental control radiator and an RCS engine quad on the exterior panel and contains the SPS engine fuel sump tank. This tank feeds the engine and is also connected by feed lines to the fuel storage tank in Sector six.

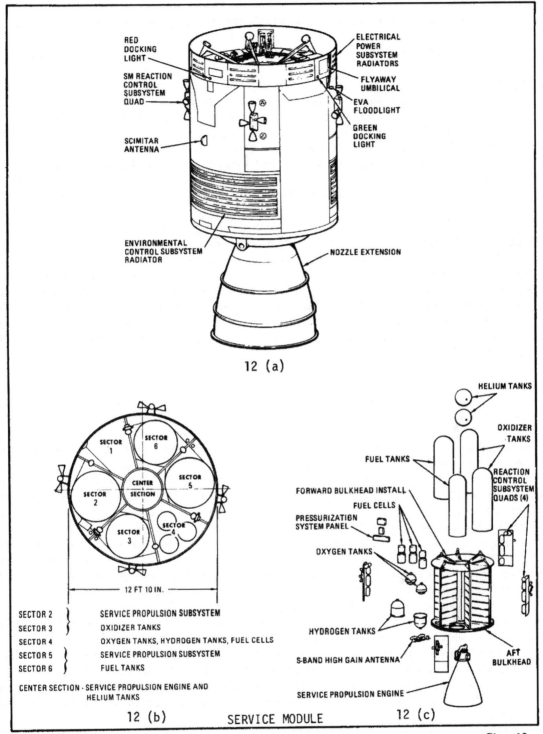

12 (a)

SECTOR 2 } SERVICE PROPULSION SUBSYSTEM
SECTOR 3 } OXIDIZER TANKS
SECTOR 4 } OXYGEN TANKS, HYDROGEN TANKS, FUEL CELLS
SECTOR 5 } SERVICE PROPULSION SUBSYSTEM
SECTOR 6 } FUEL TANKS

CENTER SECTION - SERVICE PROPULSION ENGINE AND
                 HELIUM TANKS

12 (b)   SERVICE MODULE   12 (c)

Fig. 12

Sector six has the rest of the environmental control radiator and an RCS engine quad on its exterior and contains the SPS engine fuel storage tank which feeds the fuel sump tank in Sector five.

The center section contains two helium tanks and the SPS engine. The tanks are used to provide helium pressurant for the SPS propellant tanks.

PROPULSION

The SPS engine is a restartable, non-throttleable engine which uses nitrogen tetroxide as an oxidizer and a 50-50 mixture of hydrozine and unsymmetrical dimethylhydrazine as fuel. This engine is used for major velocity changes during the mission such as midcourse corrections, lunar orbit insertion and transearth injection. The service propulsion engine responds to automatic firing commands from the guidance and navigation system or to commands from manual controls. The engine assembly is gimbal-mounted to allow engine thrust-vector alignment with the spacecraft center of mass to preclude tumbling. Thrust vector alignment control is maintained automatically by the stabilization and control system or manually by the crew.

ADDITIONAL SM SYSTEMS

In addition to the systems already described the SM has communication antennas, umbilical connections and several exterior mounted lights.

The four antennas on the outside of the SM are the S-band high-gain antenna, mounted on the aft bulkhead; two VHF omni-directional antennas, mounted on opposite sides of the module near the top; and the rendezvous radar transponder antenna, mounted in the SM fairing. The S-band high-gain antenna, used for deep space communications, is composed of four 31-inch diameter reflectors surrounding an 11-inch square reflector. At launch it is folded down parallel to the SPS engine nozzle so that it fits within the spacecraft LM adapter. After the CSM separates from the SLA the antenna is deployed at a right angle to the SM center line.

The umbilicals consist of the main plumbing and wiring connections between the CM and SM enclosed in a fairing (aluminum covering), and a "flyaway" umbilical which is connected to the launch tower. The latter supplies oxygen and nitrogen for cabin pressurization, water-glycol, electrical power from ground equipment, and purge gas.

Seven lights are mounted in the aluminum panels of the fairing. Four (one red, one green, and two amber) are used to aid the astronauts in docking, one is a floodlight which can be turned on to give astronauts visibility during extravehicular activities, one is a flashing beacon used to aid in rendezvous, and one is a spotlight used in rendezvous from 500 feet to docking with the LM.

SM/CM SEPARATION

Separation of the SM from the CM occurs shortly before entry. The sequence of events during separation is controlled automatically by two redundant Service Module Jettison Controllers(SMJC) located on the forward bulkhead of the SM. Physical separation requires severing of all the connections between the modules, transfer of electrical control, and firing of the SMRCS to increase the distance between the CM and SM.

A tenth of a second after electrical connections are deadfaced, the SMJC's send signals which fire ordnance devices to sever the three tension ties and the umbilical. The tension ties are straps which hold the CM on three of the compression pads on the SM. Linear-shaped charges in each tension-tie assembly sever the tension ties to separate the CM from the SM. At the same time, explosive charges drive guillotines through the wiring and tubing in the umbilical.

Simultaneously with the firing of the ordnance devices, the SMJC's send signals which fire the SMRCS. Roll engines are fired for five seconds to alter the SM's course from that of the CM, and the translation (thrust) engines are fired continuously until the propellant is depleted or fuel cell power is expended. These maneuvers carry the SM well away from the entry path of the CM.

COMMAND MODULE GENERAL

The Command Module (CM) (Figure 13) serves as the command, control and communications center for most of the mission. Supplemented by the SM, it provides all life support elements for three crewmen in the mission environments and for their safe return to earth's surface. It is capable of attitude control about three axes and some lateral lift translation at high velocities in earth atmosphere. It also permits LM attachment, CM/LM ingress and egress, and serves as a buoyant vessel in open ocean.

STRUCTURE

The CM consists of two basic structures joined together: the inner structure (pressure shell) and the outer structure (heat shield).

The inner structure, the pressurized crew compartment, is made of aluminum sandwich construction consisting of a welded aluminum inner skin, bonded aluminum honeycomb core and outer face

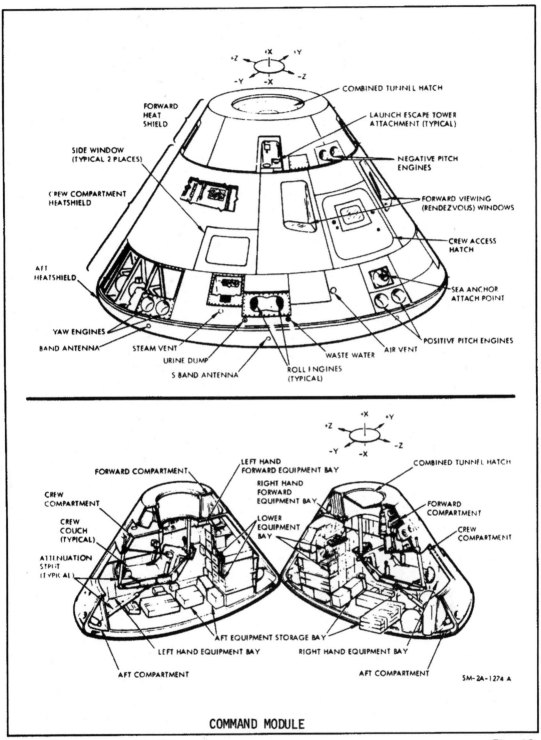

COMMAND MODULE

Fig. 13

sheet. The outer structure is basically a heat shield and is made of stainless steel brazed honeycomb brazed between steel alloy face sheets. Parts of the area between the inner and outer sheets is filled with a layer of fibrous insulation as additional heat protection.

THERMAL PROTECTION (HEAT SHIELDS)

The interior of the CM must be protected from the extremes of environment that will be encountered during a mission. The heat of launch is absorbed principally through the Boost Protective Cover (BPC), a fiberglass structure covered with cork which encloses the CM. The cork is covered with a white reflective coating. The BPC is permanently attached to the launch escape tower and is jettisoned with it.

The insulation between the inner and outer shells, plus temperature control provided by the environmental control subsystem, protects the crew and sensitive equipment in space. The principal task of the heat shield that forms the outer structure is to protect the crew during re-entry. This protection is provided by ablative heat shields, of varying thicknesses, covering the CM. The ablative material is a phenolic epoxy resin. This material turns white hot, chars, and then melts away, conducting relatively little heat to the inner structure. The heat shield has several outer coverings: a pore seal, a moisture barrier (a white reflective coating), and a silver Mylar thermal coating.

FORWARD COMPARTMENT

The forward compartment is the area around the forward (docking) tunnel. It is separated from the crew compartment by a bulkhead and covered by the forward heat shield. The compartment is divided into four 90-degree segments which contain earth landing equipment (all the parachutes, recovery antennas and beacon light, and sea recovery sling, etc.), two RCS engines, and the forward heat shield release mechanism.

The forward heat shield contains four recessed fittings into which the legs of the launch escape tower are attached. The tower legs are connected to the CM structure by frangible nuts containing small explosive charges, which separate the tower from the CM when the LES is jettisoned.

The forward heat shield is jettisoned at about 25,000 feet during return to permit deployment of the parachutes.

AFT COMPARTMENT

The aft compartment is located around the periphery of the CM at its widest part, near the aft heat shield. The aft compartment bays contain 10 RCS engines; the fuel, oxidizer, and helium tanks for the CM RCS; water tanks; the crushable ribs of the impact attenuation system; and a number of instruments. The CM-SM umbilical is also located in the aft compartment.

CREW COMPARTMENT

The crew compartment has a habitable volume of 210 cubic feet. Pressurization and temperature are maintained by the ECS. The crew compartment contains the controls and displays for operation of the spacecraft, crew couches, and all the other equipment needed by the crew. It contains two hatches, five windows, and a number of equipment bays.

EQUIPMENT BAYS

The equipment bays contain items needed by the crew for up to 14 days, as well as much of the electronics and other equipment needed for operation of the spacecraft. The bays are named according to their position with reference to the couches.

The lower equipment bay is the largest and contains most of the guidance and navigation electronics, as well as the sextant and telescope, the Command Module Computer (CMC), and a computer keyboard. Most of the telecommunications subsystem electronics are in this bay, including the five batteries, inverters, and battery charger of the electrical power subsystem. Stowage areas in the bay contain food supplies, scientific instruments, and other astronaut equipment.

The left-hand equipment bay contains key elements of the ECS. Space is provided in this bay for stowing the forward hatch when the CM and LM are docked and the tunnel between the modules is open.

The left-hand forward equipment bay also contains ECS equipment, as well as the water delivery unit and clothing storage.

The right-hand equipment bay contains waste management system controls and equipment, electrical power equipment, and a variety of electronics, including sequence controllers and signal conditioners. Food also is stored in a compartment in this bay.

The right-hand forward equipment bay is used principally for stowage and contains such items as survival kits, medical supplies, optical equipment, the LM docking target, and bio-instrumentation harness equipment.

The aft equipment bay is used for storing space suits and helmets, life vests, the fecal canister, portable life support systems (backpacks), and other equipment, and includes space for stowing the probe and drogue assembly.

HATCHES

The two CM hatches are the side hatch, used for getting in and out of the CM, and the forward hatch, used to and from the LM when the CM and LM are docked.

The side hatch is a single integrated assembly which opens outward and has primary and secondary thermal seals. The hatch normally contains a small window, but has provisions for installation of an airlock.

The latches for the side hatch are so designed that pressure exerted against the hatch serves only to increase the locking pressure of the latches.

The hatch handle mechanism also operates a mechanism which opens the access hatch in the BPC. A counterbalance assembly which consists of two nitrogen bottles and a piston assembly enables the hatch and BPC hatch to be opened easily. In space, the crew can operate the hatch easily without the counter balance and the piston cylinder and nitrogen bottle can be vented after launch. A second nitrogen bottle can be used to open the hatch after landing. The side hatch can readily be opened from the outside.

In case some deformation or other malfunction prevented the latches from engaging, three jackscrews are provided in the crew's tool set to hold the door closed.

The forward (docking) hatch is a combined pressure and ablative hatch mounted at the top of the docking tunnel. The exterior or upper side of the hatch is covered with a half-inch of insulation and a layer of aluminum foil.

This hatch has a six-point latching arrangement operated by a pump handle similar to that on the side hatch and can also be opened from the outside. It has a pressure equalization valve so that the pressure in the tunnel and that in the LM can be equalized before the hatch is removed. There are also provisions for opening the latches manually if the handle gear mechanism should fail.

WINDOWS

The CM has five windows: two side, two rendezvous, and a hatch window. The hatch window is over the center couch. The windows each consist of inner and outer panes. The inner windows are made of tempered silica glass with quarter-inch thick double panes, separated by a tenth of an inch. The outer windows are made of amorphous-fused silicon with a single pane seven tenths of an inch thick. Each pane has an antireflecting coating on the external surface and a blue-red reflective coating on the inner surface to filter out most infrared and all ultraviolet rays. The outer window glass has a softening temperature of 2800° F and a melting point of 3110°F. The inner window glass has a softening temperature of 2000° F.

Aluminum shades are provided for all windows.

IMPACT ATTENUATION

During a water impact the CM deceleration force will vary considerably depending on the shape of the waves and the dynamics of the CM's descent. A major portion of the energy (75 to 90 percent) is absorbed by the water and by deformation of the CM structure. The module's impact attenuation system reduces the forces acting on the crew to a tolerable level. The impact attenuation system is part internal and part external. The external part consists of four crushable ribs (each about four inches thick and a foot in length) installed in the aft compartment. The ribs are made of bonded laminations of corrugated aluminum which absorb energy by collapsing upon impact. The main parachutes suspend the CM at such an angle that the ribs are the first point of the module that hits the water. The internal portion of the system consists of eight struts which connect the crew couches to the CM structure. These struts absorb energy by deforming steel wire rings between an inner and an outer piston.

DISPLAYS AND CONTROLS

The Main Display Console (Figure 14) has been arranged to provide for the expected duties of crew members. These duties fall into the categories of Commander, CM Pilot, and LM Pilot, occupying the left, center, and right couches, respectively. The CM Pilot also acts as the principal navigator. Flight controls are located on the left-center and left side of the Main Display Console, opposite the Commander. These include controls for such subsystems as stabilization and control, propulsion, crew safety, earth landing, and emergency detection. One of two guidance and navigation computer panels also is located here, as are

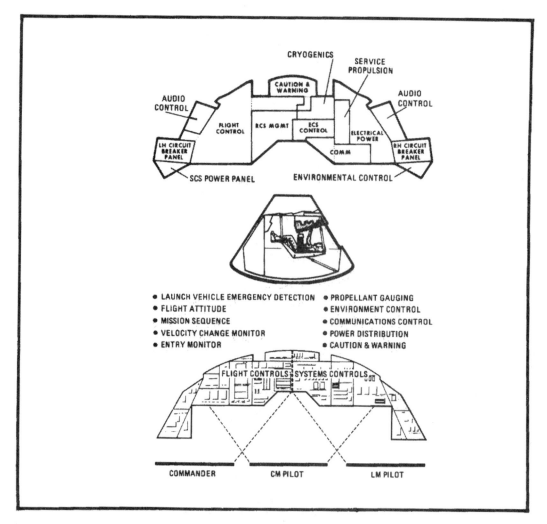

MAIN DISPLAY CONSOLE   Figure 14.

velocity, attitude, and altitude indicators.

The CM Pilot faces the center of the console, and thus can reach many of the flight controls, as well as the system controls on the right side of the console. Displays and controls directly opposite him include reaction control propellant management, caution and warning, environmental control and cryogenic storage systems.

The LM Pilot couch faces the right-center and right side of the console. Communications, electrical control, data storage, and fuel cell system components are located here, as well as service propulsion subsystem propellant managements.

All controls have been designed so they can be operated by astronauts wearing gloves. The controls are predominantly of four basic types: toggle switches, rotary switches with click-stops, thumbwheels, and push buttons. Critical switches are guarded so that they cannot be thrown inadvertently. In addition, some critical controls have locks that must be released before they can be operated.

Other displays and controls are placed throughout the cabin in the various equipment bays and on the crew couches. Most of the guidance and navigation equipment is in the lower equipment bay, at the foot of the center couch. This equipment, including the sextant and telescope, is operated by an astronaut standing and using a simple restraint system. The non-time-critical controls of the environmental control system are located in the left-hand equipment bays, while all the controls of the waste management system are on a panel in the right-hand equipment bay. The rotation and transition controllers used for attitude, thrust vector, and transition maneuvers are located on the arms of two crew couches. In addition, a rotation controller can be mounted at the navigation position in the lower equipment bay.

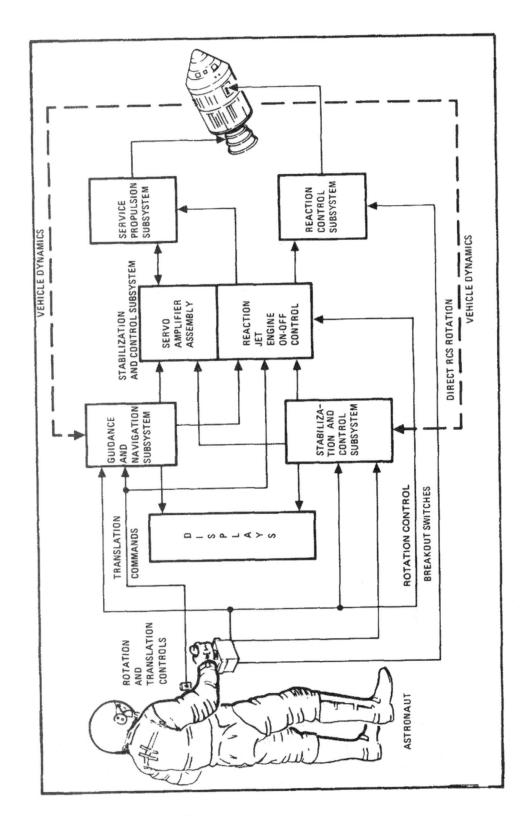

GUIDANCE AND CONTROL FUNCTIONAL FLOW

Fig. 15

Critical conditions of most spacecraft systems are monitored by a Caution And Warning System. A malfunction or out-of-tolerance condition results in illumination of a status light that identifies the abnormality. It also activates the master alarm circuit, which illuminates two master alarm lights on the Main Display Console and one in the lower equipment bay and sends an alarm tone to the astronauts' headsets. The master alarm lights and tone continue until a crewman resets the master alarm circuit. This can be done before the crewmen deal with the problem indicated. The Caution And Warning System also contains equipment to sense its own malfunctions.

GUIDANCE AND CONTROL

The Apollo spacecraft is guided and controlled by two interrelated systems (Figure 15). One is the Guidance, Navigation, and Control (GNCS) System. The other is the Stabilization and Control System (SCS).

The two systems provide rotational, line-of-flight, and rate-of-speed information. They integrate and interpret this information and convert it into commands for the spacecraft's propulsion systems.

GUIDANCE, NAVIGATION, AND CONTROL SYSTEM

Guidance and navigation is accomplished through three major elements. They are the inertial, optical, and computer systems.

The inertial subsystem senses any changes in the velocity and angle of the spacecraft and relays this information to the computer which transmits any necessary signals to the spacecraft engines.

The optical subsystem is used to obtain navigation sightings of celestial bodies and landmarks on the earth and moon. It passes this information along to the computer for guidance and control purposes.

The computer subsystem uses information from a number of sources to determine the spacecraft position and speed and, in automatic operation, to give commands for guidance and control.

STABILIZATION AND CONTROL SYSTEM

The Stabilization and Control System (SCS) operates in three ways: it determines the spacecraft's attitude (angular position); it maintains the spacecraft's attitude; it controls the direction of thrust of the service propulsion engine.

Both the GNCS and SCS are used by the computer in the Command Module to provide automatic control of the Spacecraft. Manual control of the spacecraft attitude and thrust is provided mainly through the SCS equipment.

Spacecraft Attitude

The Flight Director Attitude Indicators (FDAI) on the main console show the total angular position, attitude errors and their rates of change. One of the sources of total attitude information is the stable platform of the Inertial Measurement Unit (IMU). The second source is a Gyro Display Coupler (GDC) which gives a reading of the spacecraft's actual attitudes as compared with an attitude selected by the crew.

Information about attitude error also is obtained by comparison of the IMU gimbal angles with computer reference angles. Another source of this information is gyro assembly No. 1, which senses any spacecraft rotation about any of the three axes.

Total attitude information goes to the Command Module Computer (CMC) as well as to the FDAI's on the console.

Attitude Control

If a specific attitude or orientation is desired, attitude error signals are sent to the reaction jet engine control assembly. Then the proper reaction jet automatically fires in the direction necessary to return the spacecraft to the desired position.

Thrust Control

The CMC provides primary control of thrust. The flight crew pre-sets thrusting and spacecraft data into the computer by means of the display keyboard. The forthcoming commands include time and duration of thrust. Accelerometers sense the amount of change in velocity obtained by the thrust.

Thrust direction control is required because of center of gravity shifts caused by depletion of propellants in service propulsion tanks. This control is accomplished through electromechanical actuators which position the gimbaled service propulsion engine. Automatic control commands may originate in either

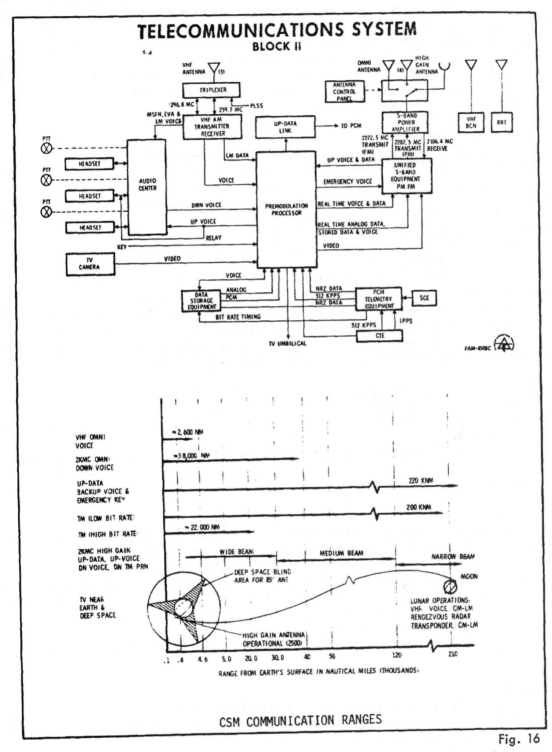

TELECOMMUNICATIONS SYSTEM
BLOCK II

CSM COMMUNICATION RANGES

Fig. 16

the guidance and navigation subsystem or the SCS. There is also provision for manual controls.

TELECOMMUNICATIONS

The telecommunications system (Figure 16) provides voice, television, telemetry, and tracking and ranging communications between the spacecraft and earth, between the CM and LM, and between the spacecraft and astronauts wearing the Portable Life Support System (PLSS). It also provides communications among the astronauts in the spacecraft and includes the central timing equipment for synchronization of other equipment and correlation of telemetry equipment.

For convenience, the telecommunications subsystem can be divided into four areas intercommunications (voice), data, radio frequency equipment, and antennas.

INTERCOMMUNICATIONS

The astronauts' headsets are used for all voice communications. Each astronaut has an audio control panel on the main display console which enables him to control what comes into his headset and where he will send his voice.

The three headsets and audio control panels are connected to three identical audio center modules.

The audio center is the assimilation and distribution point for all spacecraft voice signals. The audio signals can be routed from the center to the appropriate transmitter or receiver, the Launch Control Center (for pre-launch checkout), the recovery forces intercom, or voice tape recorders.

Two methods of voice transmission and reception are possible: the VHF/AM transmitter-receiver and the S-band transmitter and receiver.

The VHF/AM equipment is used for voice communications with the Manned Space Flight Network during launch, ascent, and near-earth phases of a mission. The S-band equipment is used during both near-earth and deep-space phases of a mission. When communications with earth are not possible, a limited number of audio signals can be stored on tape for later transmission.

DATA

The spacecraft structure and subsystems contain sensors which gather data on their status and performance. Biomedical, TV, and timing data also are gathered. These various forms of data are assimilated into the data system, processed, and then transmitted to the ground. Some data from the operational systems, and some voice communications, may be stored for later transmission or for recovery after landing. Stored data can be transmitted to the ground simultaneously with voice or real-time data.

RADIO FREQUENCY EQUIPMENT

The radio frequency equipment is the means by which voice information, telemetry data, and ranging and tracking information are transmitted and received. The equipment consists of two VHF/AM transceivers in one unit, the unified S-band equipment (primary and secondary transponders and an FM transmitter), primary and secondary S-band power amplifiers (in one unit), a VHF beacon, an X-band transponder (for rendezvous radar), and the premodulation processor.

The equipment provides for voice transfer between the CM and the ground, between the CM and LM, between the CM and extravehicular astronauts, and between the CM and recovery forces. Telemetry can be transferred between the CM and the ground, from the LM to the CM and then to the ground, and from extravehicular astronauts to the CM and then to the ground. Ranging information consists of pseudo-random noise and double-doppler ranging signals from the ground to the CM and back to the ground, and, of X-band radar signals from the LM to the CM and back to the LM. The VHF beacon equipment emits a 2-second signal every five seconds for line-of-sight direction finding to aid recovery forces in locating the CM after landing.

ANTENNAS

There are nine antennas on the CSM, not counting the rendezvous radar antenna which is an integral part of the rendezvous radar transponder. These antennas are shown in Figure 17.

These antennas can be divided into four groups: VHF, S-band, recovery, and beacon. The two VHF antennas (called scimitars because of their shape) are omni-directional and are mounted 180 degrees apart on the SM. There are five S-band antennas, one mounted at the bottom of the SM and four located 90 degrees apart around the CM. The four smaller surface-mounted S-band antennas are used at near-earth ranges and deep-space backup. The high-gain antenna is deployable after CSM/SLA separation. It can be steered through a gimbal system and is the principal antenna for deep-space communications. The four S-band antennas on the CM are mounted flush with the surface of the CM and are used for S-band communications during near-earth phases of the mission, as well as for a backup in deep space. The two VHF recovery antennas are located in the forward compartment of the CM, and are deployed automatically shortly after the main parachutes deploy. One of these antennas also is connected to the VHF recovery beacon.

ENVIRONMENTAL CONTROL SYSTEM

The Environmental Control System (ECS) provides a controlled environment for three astronauts

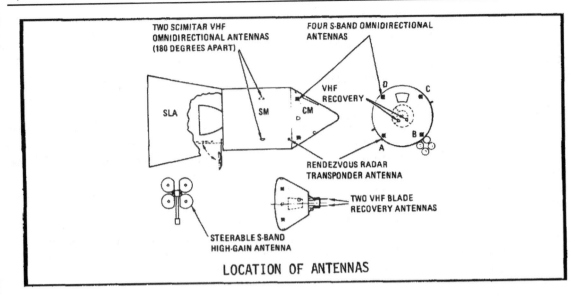

LOCATION OF ANTENNAS

Fig. 17

for up to 14 days. For normal conditions, this environment includes a pressurized cabin (five pounds per square inch), a 100-percent oxygen atmosphere, and a cabin temperature of 70 to 75 degrees fahrenheit. For use during critical mission phases and for emergencies, the subsystem provides a pressurized suit circuit.

The system provides oxygen and hot and cold water, removes carbon dioxide and odors from the CM cabin, provides for venting of waste, and dissipates excessive heat from the cabin and from operating electronic equipment. It is designed so that a minimum amount of crew time is needed for its normal operation. The main unit contains the coolant control panel, water chiller, two water-glycol evaporators, carbon dioxide-odor absorber canisters, suit heat exchanger, water separator, and compressors. The oxygen surge tank, water glycol pump package and reservoir, and control panels for oxygen and water are adjacent to the unit. The system is concerned with three major elements: oxygen, water, and coolant (water-glycol). All three are interrelated and intermingled with other systems. These three elements provide the major functions of spacecraft atmosphere, thermal control, and water management through four major subsystems: oxygen, pressure suit circuit, water, and water-glycol. A fifth subsystem, post-landing ventilation, also is part of the environmental control system, providing outside air for breathing and cooling prior to hatch opening.

The CM cabin atmosphere is 60 percent oxygen and 40 percent nitrogen on the launch pad to reduce fire hazard. The mixed atmosphere supplied by ground equipment will gradually be changed to pure oxygen after launch as the environmental control system maintains pressure and replenishes the cabin atmosphere. During pre-launch and initial orbital operation, the suit circuit supplies pure oxygen at a flow rate slightly more than is needed for breathing and suit leakage. This results in the suit being pressurized slightly above cabin pressure, which prevents cabin gases from entering and contaminating the suit circuit. The excess oxygen in the suit circuit is vented into the cabin.

Spacecraft heating and cooling is performed through two water-glycol coolant loops. The water-glycol, initially cooled through ground equipment, is pumped through the primary loop to cool operating electric and electronic equipment, the space suits, and the cabin heat exchangers. The water-glycol also is circulated through a reservoir in the CM to provide a heat sink during ascent.

CM REACTION CONTROL SYSTEM

The CM Reaction Control System (RCS) is used after CM/SM separation and for certain abort modes. It provides three-axis attitude control to orient and maintain the CM in the proper entry attitude.

The system consists of two independent, redundant systems. The two systems can operate in tandem; however, one can provide all the impulse needed for the entry maneuvers, and normally only one is used.

The 12 engines of the system are located outside the crew compartment of the CM, ten in the aft compartment and two in the forward compartment. Each engine produces approximately 93 pounds of thrust.

Operation of the CM RCS is similar to that of the SM RCS. The fuel is monomethyl hydrazine and the oxidizer is nitrogen tetroxide. Helium is used for pressurization. Each of the redundant CM systems contains one fuel and one oxidizer tank similar to the fuel and oxidizer tanks of the SM system. Each CM system has one helium tank

ELECTRICAL POWER SYSTEM

The Electrical Power System (EPS) provides electrical energy sources, power generation and control, power conversion and conditioning, and power distribution to the spacecraft throughout the mission. The EPS also furnishes drinking water to the astronauts as a by-product of the fuel cells. The primary source of electrical power is the fuel cells mounted in the SM. Each cell consists of a hydrogen compartment, an oxygen compartment, and two electrodes.

The cryogenic gas storage system, also located in the SM, supplies the hydrogen and oxygen used in the fuel cell power plants, as well as the oxygen used in the ECS.

Three silver oxide-zinc storage batteries supply power to the CM during entry and after landing, provide power for sequence controllers, and supplement the fuel cells during periods of peak power demand. These batteries are located in the CM lower equipment bay. A battery charger is located in the same bay to assure a full charge prior to entry.

Two other silver oxide-zinc batteries, independent of and completely isolated from the rest of the dc power system, are used to supply power for explosive devices for CM/SM separation, parachute deployment and separation, third-stage separation, launch escape system tower separation, and other pyrotechnic uses.

EMERGENCY DETECTION SYSTEM

The Emergency Detection System (EDS) monitors critical conditions of launch vehicle powered flight. Emergency conditions are displayed to the crew on the main display console to indicate a necessity for abort. The system includes provisions for a crew-initiated abort with the use of the LES or with the SPS after tower jettison. The crew can initiate an abort separation from the LV from prior to lift-off until the planned separation time. A capability also exists for commanding early staging of the S-IVB from the S-II stage when necessary. Also included in the system are provisions for an automatic abort in case of the following time-critical conditions:

1. Loss of thrust on two or more engines on the first stage of the Launch Vehicle.
2. Excessive vehicle angular rates in any of the pitch, yaw, or roll planes.
3. Loss of "hotwire" continuity from SM to IU.

The EDS will automatically initiate an abort signal when two or more first-stage engines are out or when Launch Vehicle excessive rates are sensed by gyros in the Instrument Unit. The abort signals are sent to the master events sequence controller, which initiates the abort sequence. The engine lights on the Main Display Console provide the following information to the crew: ignition, cutoff, engine below pre-specified thrust level, and physical stage separation. A yellow "S-II Sep" light will illuminate at second-stage first-plane separation and will extinguish at second-plane separation. A high-intensity, red "ABORT" light is illuminated if an abort is requested by the Launch Control Center for a pad abort or an abort during lift-off via updata link. The "ABORT" light also can be illuminated after lift-off by the Range Safety Officer, or by the Mission Control Center via the updata link from the Manned Space Flight Network.

EARTH LANDING SYSTEM

The Earth Landing System (ELS) provides a safe landing for the astronauts and the CM. Several recovery aids which are activated after splashdown are part of the system. Operation normally is automatic, timed and activated by the sequential control system. All automatic functions can be backed up manually. Components and their locations are shown in Figure 18.

For normal entry, about 1.5 seconds after forward heat shield jettison, the two drogue parachutes are deployed to orient the CM properly and to provide initial deceleration. At about 10,000 feet, the drogue parachutes are released and the three pilot parachutes are deployed; these pull the main parachutes from the forward section of the CM. The main parachutes initially open partially (reefed) for ten seconds, to limit

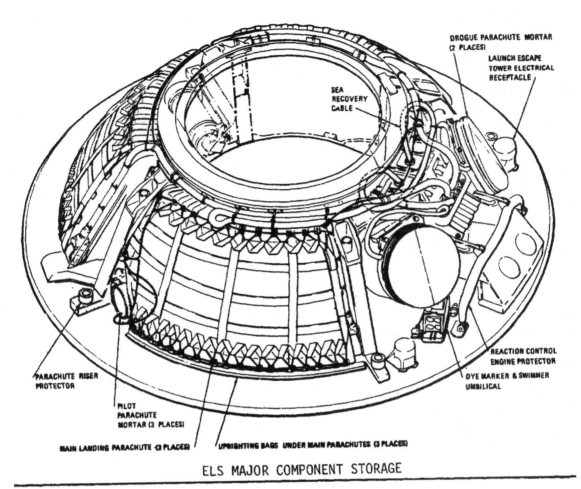

DROGUE PARACHUTE MORTAR
(2 PLACES)

LAUNCH ESCAPE
TOWER ELECTRICAL
RECEPTACLE

SEA
RECOVERY
CABLE

PARACHUTE RISER
PROTECTOR

PILOT
PARACHUTE
MORTAR (3 PLACES)

MAIN LANDING PARACHUTE (3 PLACES)

UPRIGHTING BAGS UNDER MAIN PARACHUTES (3 PLACES)

REACTION CONTROL
ENGINE PROTECTOR

DYE MARKER & SWIMMER
UMBILICAL

ELS MAJOR COMPONENT STORAGE

Fig. 18

deceleration prior to full-diameter deployment. The main parachutes hang the CM at an angle of 27.5 degrees to decrease impact loads at touchdown.

After splashdown, the main parachutes are released and the recovery aid subsystem is set in operation by the crew. The subsystem consists of an uprighting system, swimmer's umbilical cable, a sea dye marker, a flashing beacon, and a VHF beacon transmitter. A sea recovery sling of steel cable also is provided to lift the CM aboard a recovery ship.

The two VHF recovery antennas are located in the forward compartment with the parachutes. They are deployed automatically eight-seconds after the main parachutes. One of them is connected to the beacon transmitter, which emits a two-second signal every five-seconds to aid recovery forces in locating the CM. The other is connected to the VHF/AM transmitter and receiver to provide voice communications between the crew and recovery forces.

Three inflatable uprighting bags, stowed under the main parachutes, are available for uprighting the CM should it stabilize in an inverted floating position after splashdown.

Automatic operation of the Earth Landing System is provided by the event sequencing system located in the right-hand equipment bay of the CM. The system contains the barometric pressure switches, time delays, and relays necessary to control jettisoning of the heat shield and the deployment of the parachutes.

CREW PROVISIONS
APPAREL

The number of items a crewman wears varies during a mission. There are three basic conditions: unsuited, suited, and extravehicular. Unsuited, the crewman breathes the cabin oxygen and wears a bio-instrumentation harness, a communication soft hat, a constant-wear garment, flight coveralls, and booties.

Under the space suit, the astronaut wears the bio-instrumentation harness, communications soft hat, and constant-wear garment. The extravehicular outfit, designed primarily for wear during lunar exploration, includes the bio-instrumentation harness, a fecal containment system, a liquid-cooled garment, the communication soft hat, the space suit, a portable life support system (backpack), an oxygen purge system, a thermal/ meteoroid garment, and an extravehicular visor.

The bio-instrumentation harness has sensors, signal conditioners, a belt, and wire signal carriers. These monitor the crewman's physical condition and the information is telemetered to the ground.

The constant-wear garment is an undergarment for the suit and flight coveralls. Made of porous cotton cloth, it is a one-piece, short-sleeved garment with feet similar to those on long underwear. It is zippered from waist to neck and has openings front and rear for personal hygiene. Pockets at the ankles, thighs, and chest hold passive radiation dosimeters. Spare garments are stowed on the aft bulkhead.

The flight coverall is the basic outer garment for unsuited operations. It is a two piece, Teflon-cloth garment with pockets on the shins and thighs for personal equipment.

The communication soft hat is worn when suited. It has two earphones and two microphones, with voice tubes on two mounts that fit over the ears. An electrical cable runs from the hat to the communications cable. A lightweight headset is worn when crewmen are not in their suits.

Booties worn with the flight coveralls are made of Beta cloth, with Velcro hook material bonded to the soles. During weightlessness, the Velcro hook engages Velcro pile patches attached to the floor to hold the crewman in place.

SPACE SUIT

The space suit protects the astronaut from the hostile environment of space. It provides atmosphere for breathing and pressurization, protects him from heat, cold, and micro-meteroroids, and contains a communications link.

The suit is worn by the astronauts during all critical phases of the mission, during periods when the CM is unpressurized, and during all operations outside the CM and LM, whether in space or on the Moon.

The suit system must provide an artificial atmosphere (100-percent oxygen for breathing and for pressurization to 3.7 psi), adequate mobility, micro-meteorite and visual protective systems, and the ability to operate on the lunar surface for periods of three hours. Design of the Apollo spacecraft and suits will permit the crew to operate — with certain restraints — in a decompressed cabin for periods as long as 115 hours.

The complete space suit is called the Pressure Garment Assembly (PGA). It is composed of a number of items assembled into two configurations: extravehicular (for outside the spacecraft) and intravehicular. The addition of the backpack to the extravehicular space suit makes up the Extravehicular Mobility Unit (EMU). The backpack (called the Portable Life Support System PLSS) supplies oxygen, electrical power, communications, and liquid cooling.

The intravehicular space suit consists of : fecal containment subsystem, constant-wear garment, biomedical belt, urine collection transfer assembly, torso limb suit, integrated thermal/ micro-meteoroid garment, pressure helmet, pressure glove, and communications carrier.

In the extravehicular configuration, the constant wear garment is replaced by the liquid-cooling garment and four items are added to the intravehicular suit: extravehicular visor, extravehicular glove, lunar overshoe, and a cover which fits over umbilical connections on the front of the suit.

The pressure suit is a white garment that weighs about 60 pounds with the integrated thermal/ micro-meteoroid garment. The latter weighs about 19 pounds.

The basic components of the suit, or PGA, are the torso limb suit, the pressure helmet, the pressure glove, the integrated thermal/ micro-meteoroid garment, and the extravehicular glove.

A cable and hose assembly connects the space suits to the spacecraft. The cable provides a communications capability and the two hoses carry oxygen. There are three fluorel-coated hose assemblies. The communications cable assembly consists of a cable and control head. The cables run next to the two hoses of the assembly.

FOOD AND WATER

Food supplies are designed to supply each astronaut with a balanced diet of approximately 2800 calories a day. The food is either freeze-dried or concentrated and is carried in vacuum-packaged plastic bags. Each bag of freeze-dried food has a oneway valve through which water is inserted and a second valve through which food passes. Concentrated food is packaged in bite-size units and needs no reconstitution. Several bags are packaged together to make one meal bag. The meal bags have red, white, and blue dots to identify them for each crewman, as well as labels to identify them by day and meal.

The food is reconstituted by adding hot or cold water through the one-way valve. The astronaut kneads the bag and then cuts the neck of the bag and squeezes the food into his mouth.

Drinking water comes from the water chiller to two outlets: the water meter dispenser, and the food preparation unit. The dispenser has an aluminum mounting bracket, a 72-inch coiled hose, and a dispensing valve unit in the form of a button-actuated pistol. The pistol barrel is placed in the mouth and the button is pushed for each half-ounce of water. The meter records the amount of water drunk. A valve is provided to shut off the system in case the dispenser develops a leak or malfunction.

Food preparation water is dispensed from a unit which has hot (150°F) and cold (50°F) water. Cold water comes directly to the unit from the water chiller. Hot water is accumulated in a 38-ounce tank which contains three heaters that keep the water at 150° F.

COUCHES AND RESTRAINTS

The astronaut couches are individually adjustable units made of hollow steel tubing and covered with a heavy, fireproof, fiberglass cloth. The couches rest on a head beam and two side-stabilizer beams supported by eight attenuator struts (two each for the Y and Z axes and four for the X axis) which absorb the impact of landing.

These couches support the crewmen during acceleration and deceleration, position the crewman at their duty stations, and provide support for translation and rotation hand controls, lights, and other equipment. A lap belt and shoulder straps are attached to the couches.

The couches can be folded or adjusted into a number of seat positions. The one used most is the 85-degree position assumed for launch, orbit entry, and landing. The 170 degree (flat-out) position is used primarily for the center couch, so that crewmen can move into the lower equipment bay. The armrests on either side of the center couch can be folded footward so the astronauts from the two outside couches can slide over easily. The hip pan of the center couch can be disconnected and the couch can be pivoted around the head beam and laid on the aft bulkhead floor of the CM. This provides both room for the astronauts to stand and easier access to the side hatch for extravehicular activity.

Two armrests are attached to the back pan of the left couch and two armrests are attached to the right couch. The center couch has no armrests. The translation and rotation controls can be mounted to any of the four armrests. A support at the end of each armrest rotates 100 degrees to provide proper tilt for the controls. The couch seat pan and leg pan are formed of framing and cloth, and the foot pan is all steel. The foot pan contains a boot restraint device which engages the boot heel and holds it in place.

The couch restraint harness consists of a lap belt and two shoulder straps which connect to the lap belt at the buckle. The shoulder straps connect to the shoulder beam of the couch.

Other restraints in the CM include handholds, a hand bar, hand straps, and patches of Velcro which hold crewmen when they wear sandals.

The astronauts sleep in bags under the left and right couches with heads toward the hatch. The two sleeping bags are lightweight Beta fabric 64 inches long, with zipper openings for the torso and 7-inch diameter neck openings. They are supported by two longitudinal straps that attach to storage boxes in the lower equipment bay and to the CM inner structure. The astronauts sleep in the bags when unsuited and restrained on top of the bags when they have space suits on.

HYGIENE EQUIPMENT

Hygiene equipment includes wet and dry cloths for cleaning, towels, a toothbrush, and the waste management system.

The waste management system controls and disposes of waste solids, liquids, and gases. The major portion of the system is in the right-hand equipment bay. The system stores feces, removes odors, dumps urine overboard, and removes urine from the space suit.

OPERATIONAL AIDS

These include data files, tools, worksheet, cameras, fire extinguishers, oxygen masks, medical supplies, and waste bags.

The CM has one fire extinguisher, located adjacent to the left-hand and lower equipment bays. The extinguisher weighs about eight pounds. The extinguishing agent is an aqueous gel expelled in two cubic feet of foam for approximately 30 seconds at high pressure. Fire ports are located at various panels so that the extinguisher's nozzle can be inserted to put out a fire behind the panel.

Oxygen masks are provided for each astronaut in case of smoke, toxic gas, or other hostile atmosphere in the cabin while the astronauts are out of their suits. Oxygen is supplied through a flexible hose from the emergency oxygen/ repressurization unit in the upper equipment bay.

Medical supplies are contained in an emergency medical kit, about 7 x 5 x 5 inches, which is stored in the lower equipment bay. It contains oral drugs and pills (pain capsules, stimulant, antibiotic, motion sickness, diarrhea, decongestant, and aspirin), injectable drugs (for pain and motion sickness), bandages, topical agents (first-aid cream, sun cream, and an antibiotic ointment), and eye drops.

SURVIVAL EQUIPMENT

Survival equipment, intended for use in an emergency after earth landing, is stowed in two rucksacks in the right-hand forward equipment bay.

One of the rucksacks contains a three-man rubber life raft with an inflation assembly, carbon-dioxide cylinder, a sea anchor, dye marker, and a sunbonnet for each crewman.

The other rucksack contains a beacon transceiver, survival lights, desalter kits, machete, sun glasses, water cans, and a medical kit.

The survival medical kit contains the same type of supplies as the emergency medical kit: six bandages, six injectors, 30 tablets, and one tube of all-purpose ointment.

MISCELLANEOUS EQUIPMENT

Each crewman is provided a toothbrush, wet and dry cleansing cloths, ingestible toothpaste, a 64-cubic inch container for personal items, and a two-compartment, temporary storage bag.

A special tool kit is provided which also contains three jack screws for contingency hatch closure.

LAUNCH COMPLEX GENERAL

Launch Complex 39 (LC-39), located at Kennedy Space Center, Florida, is the facility provided for the assembly, checkout, and launch of the Apollo/Saturn V Space Vehicle. Assembly and checkout of the vehicle is accomplished on a Mobile Launcher, in the controlled environment of the Vehicle Assembly Building. The space vehicle and the Mobile Launcher are then moved, as a unit, by the Crawler-Transporter to the launch site. The major elements of the launch complex shown in Figure 19 are the Vehicle Assembly Building (VAB), the Launch Control Center (LCC), the Mobile Launcher (ML), the Crawler-Transporter (C/T), the crawlerway, the Mobile Service Structure (MSS), and the launch pad.

LC-39 FACILITIES AND EQUIPMENT
VEHICLE ASSEMBLY BUILDING

The VAB provides a protected environment for receipt and checkout of the propulsion stages and IU, erection of the vehicle stages and spacecraft in a vertical position on the ML, and integrated checkout of the assembled space vehicle.

The VAB, as shown in Figure 20, is a totally-enclosed structure covering eight acres of ground. It is a structural steel building approximately 525 feet high, 518 feet wide, and 716 feet long.

The principal operational elements of the VAB are the low bay and high bay areas. A 92-foot wide transfer aisle extends through the length of the VAB and divides the low and high bay areas into equal segments (Figure 20).

The low bay area provides the facilities for receiving, uncrating, checkout, and preparation of the S-II stage, S-IVB stages and the IU.

The high bay area provides the facilities for erection and checkout of the S-IC stage; mating and erection operations of the S-II stage; S-IVB stage, IU, and spacecraft; and integrated checkout of the assembled space vehicle. The high bay area contains four checkout bays, each capable of accommodating a fully-assembled Apollo/Saturn V Space Vehicle.

Launch Control Center

The LCC, Figure 20, serves as the focal point for overall direction, control, and monitoring of space vehicle checkout and launch. The LCC is located adjacent to the VAB and at a sufficient distance from the

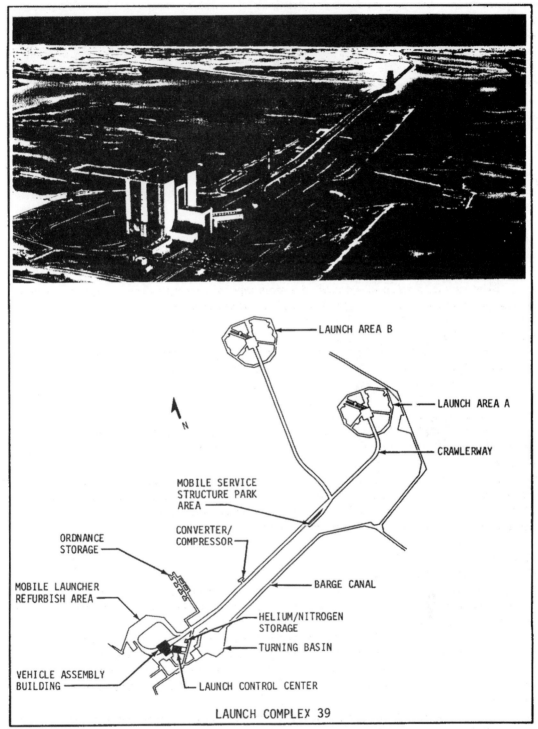

LAUNCH AREA B

LAUNCH AREA A

CRAWLERWAY

MOBILE SERVICE
STRUCTURE PARK
AREA

ORDNANCE
STORAGE

CONVERTER/
COMPRESSOR

BARGE CANAL

MOBILE LAUNCHER
REFURBISH AREA

HELIUM/NITROGEN
STORAGE

TURNING BASIN

VEHICLE ASSEMBLY
BUILDING

LAUNCH CONTROL CENTER

LAUNCH COMPLEX 39

Fig. 19

launch pad (three miles) to permit the safe viewing of lift-off without requiring site hardening.

The LCC is a four-story structure. The ground floor is devoted to service and support functions. The second floor houses telemetry and tracking equipment, in addition to instrumentation and data reduction facilities.

The third floor is divided into four separate but similar control areas, each containing a firing room, a computer room, a mission control room, a test conductor platform area, a visitor gallery, and offices. The four firing rooms, one for each high bay in the VAB, contain control, monitoring, and display equipment for

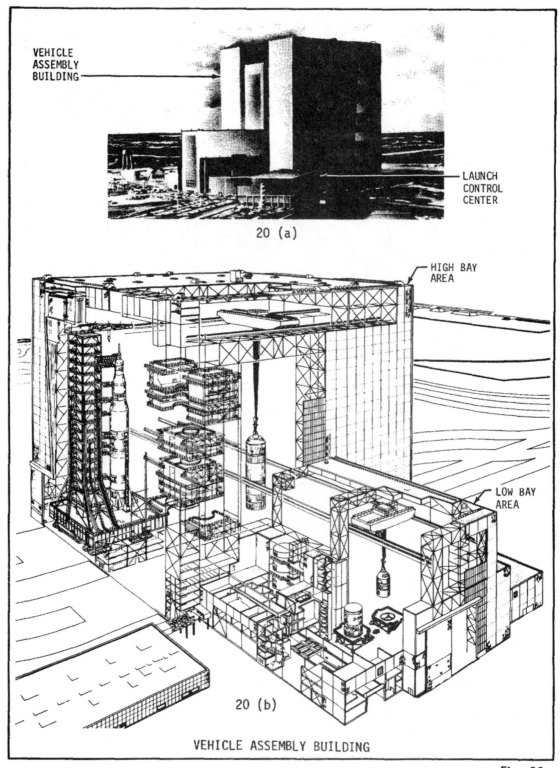

VEHICLE
ASSEMBLY
BUILDING

LAUNCH
CONTROL
CENTER

20 (a)

HIGH BAY
AREA

LOW BAY
AREA

20 (b)

VEHICLE ASSEMBLY BUILDING

Fig. 20

automatic vehicle checkout and launch.

The display rooms, offices, Launch Information Exchange Facility (LIEF) rooms, and mechanical equipment are located on the fourth floor.

The power demands in this area are large and are supplied by two separate systems, industrial and instrumentation. This division between power systems is designed to protect the instrumentation power

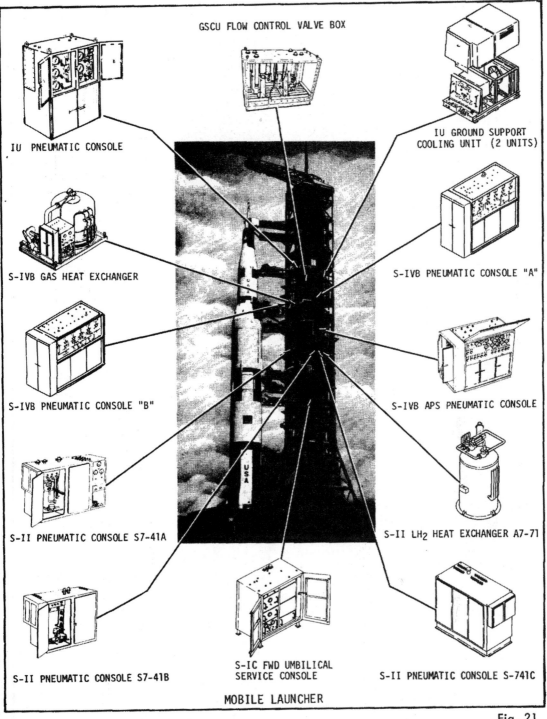

GSCU FLOW CONTROL VALVE BOX

IU PNEUMATIC CONSOLE

IU GROUND SUPPORT
COOLING UNIT (2 UNITS)

S-IVB GAS HEAT EXCHANGER

S-IVB PNEUMATIC CONSOLE "A"

S-IVB PNEUMATIC CONSOLE "B"

S-IVB APS PNEUMATIC CONSOLE

S-II PNEUMATIC CONSOLE S7-41A

S-II LH$_2$ HEAT EXCHANGER A7-71

S-II PNEUMATIC CONSOLE S7-41B

S-IC FWD UMBILICAL
SERVICE CONSOLE

S-II PNEUMATIC CONSOLE S-741C

MOBILE LAUNCHER

Fig. 21

system from the adverse effects of switching transients, large cycling loads, and intermittent motor starting loads. Communication and signal cable troughs extend from the LCC via the enclosed bridge to each ML location in the VAB high bay area. Cableways also connect to the ML refurbishing area and to the Pad Terminal Connection Room (PTCR) at the launch pad. Antennas on the roof provide an RF link to the launch pads and other facilities at KSC.

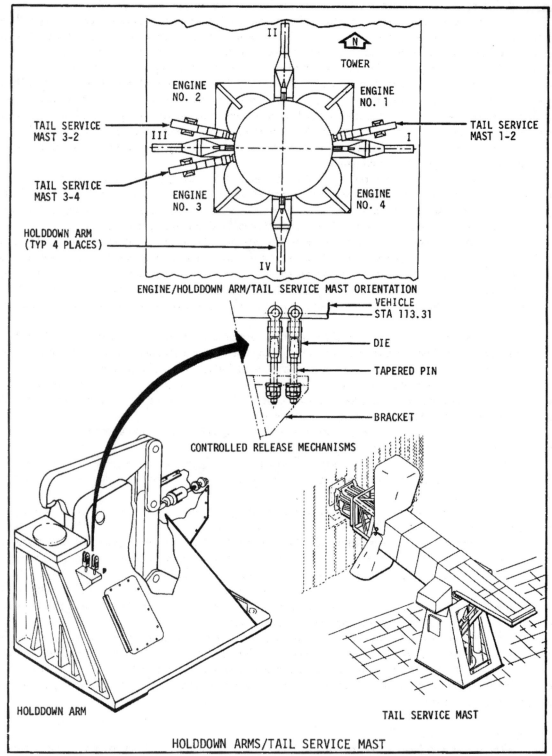

ENGINE/HOLDDOWN ARM/TAIL SERVICE MAST ORIENTATION

CONTROLLED RELEASE MECHANISMS

HOLDDOWN ARM

TAIL SERVICE MAST

HOLDDOWN ARMS/TAIL SERVICE MAST

Fig. 22

Mobile Launcher

The ML (Figure 21) is a transportable steel structure which, with the C/T, provides the capability to move the erected vehicle to the launch pad. The ML is divided into two functional areas, the launcher base and the umbilical tower. The launcher base is the platform on which a Saturn V vehicle is assembled in the vertical position, transported to a launch site, and launched. The umbilical tower provides access to all important levels of the vehicle during assembly, checkout, and servicing. The equipment used in the servicing, checkout, and launch is installed throughout both the base and tower sections of the ML.

The launcher base is a steel structure 25 feet high, 160 feet long, and 135 feet wide. The upper deck, designated level 0, contains, in addition to the umbilical tower, the four hold-down arms and the three tail service masts. There is a 45-foot square opening through the ML base for first stage exhaust.

The base has provisions for attachment to the C/T, six launcher-to-ground mount mechanisms, and four extensible support columns. All electrical/mechanical interfaces between vehicle systems and the VAB or the launch site are located through or adjacent to the base structure. The base houses such items as the computer systems test sets, digital propellant loading equipment, hydraulic test sets, propellant and pneumatic lines, air conditioning and ventilating systems, electrical power systems, and water systems.

Fueling operations at the launch area require that the compartments within the structure be pressurized with a supply of uncontaminated air.

The primary electrical power supplied to the ML is divided into four separate services: instrumentation, industrial, in-transit, and emergency. Emergency power is supplied by a diesel-driven generator located in the ground facilities. It is used for obstruction lights, emergency lighting, and for one tower elevator. Water is supplied to the ML for fire, industrial, and domestic purposes.

The umbilical tower is an open steel structure, 380 feet high, which provides the support for eight umbilical service arms, one access arm, 18 work and access platforms, distribution equipment for the propellant, pneumatic, electrical, and instrumentation subsystems, and other ground support equipment. Two high-speed elevators service 18 landings, from level A of the base to the 340-foot tower level. The structure is topped by a 25-ton hammerhead crane. Remote control of the crane is possible from numerous locations on the ML.

The four holddown arms (Figure 22) are mounted on the ML deck, 90° apart around the vehicle base. They position and hold the vehicle on the ML during the VAB checkout, movement to the pad, and pad checkout. The vehicle base is held with a pre-loaded force of 700,000 pounds at each arm.

At engine ignition, the vehicle is restrained until proper engine thrust is achieved. The unlatching interval for the four arms should not exceed 0.050 seconds. If any of the separators fail to operate in 0.180 seconds, release is effected by detonating an explosive nut link.

At launch, the holddown arms quickly release, but the vehicle is prevented from accelerating too rapidly by the controlled-release mechanisms (Figure 22).

Each controlled-release mechanism basically consists of a tapered pin inserted in a die which is coupled to the vehicle. Upon vehicle release, the tapered pin is drawn through the die during the first six inches of vehicle travel.

There are provisions for as many as 16 mechanisms per vehicle. The precise number is determined on a mission basis.

The three Tail Service Mast (TSM) assemblies (Figure 22), support service lines to the S-IC stage and provide a means for rapid retraction at vehicle lift-off. The TSM assemblies are located on level 0 of the ML base. Each TSM is a counterbalanced structure which is pneumatically/electrically controlled and hydraulically operated. Retraction of the umbilical carrier and vertical rotation of the mast is accomplished simultaneously to ensure no physical contact between the vehicle and mast. The carrier is protected by a clam-shell hood which is closed by a separate hydraulic system as the mast rotates.

The nine service arms provide access to the launch vehicle and support the service lines that are required to sustain the vehicle, as described in Figure 23. The service arms are designated as either preflight or in-flight arms. The preflight arms are retracted and locked against the umbilical tower prior to lift-off. The in-flight arms retract at vehicle lift-off. Carrier withdrawal and arm retraction is accomplished by pneumatic and/or hydraulic systems.

Launch Pad

The launch pad (Figure 24) provides a stable foundation for the ML during Apollo/ Saturn V launch and pre-launch operations and an interface to the ML for ML and vehicle systems. There are presently two pads at LC-39 located approximately three miles from the VAB area. Each launch site is approximately 3000 feet across.

The launch pad is a cellular, reinforced concrete structure with a top elevation of 42 feet above grade elevation.

Located within the fill under the west side of the structure (Figure 25) is a two-story concrete building to house environmental control and pad terminal connection equipment. On the east side of the

① S-IC Intertank (preflight). Provides lox fill and drain interfaces. Umbilical withdrawal by pneumatically driven compound parallel linkage device. Arm may be reconnected to vehicle from LCC. Retract time is 8 seconds. Reconnect time is approximately 5 minutes.

② S-IC Forward (preflight). Provides pneumatic, electrical, and air-conditioning interfaces. Umbilical withdrawal by pneumatic disconnect in conjunction with pneumatically driven block and tackle/lanyard device. Secondary mechanical system. Retracted at T-20 seconds. Retract time is 8 seconds.

③ S-II Aft (preflight). Provides access to vehicle. Arm retracted prior to liftoff as required.

④ S-II Intermediate (in-flight). Provides LH_2 and lox transfer, vent line, pneumatic, instrument cooling, electrical, and air-conditioning interfaces. Umbilical withdrawal systems same as S-IVB Forward with addition of a pneumatic cylinder actuated lanyard system. This system operates if primary withdrawal system fails. Retract time is 6.4 seconds (max).

⑤ S-II Forward (in-flight). Provides GH_2 vent, electrical, and pneumatic interfaces. Umbilical withdrawal systems same as S-IVB Forward. Retract time is 7.4 seconds (max).

⑥ S-IVB Aft (in-flight). Provides LH_2 and lox transfer, electrical, pneumatic, and air-conditioning interfaces. Umbilical withdrawal systems same as S-IVB Forward. Also equipped with line handling device. Retract time is 7.7 seconds (max).

⑦ S-IVB Forward (in-flight). Provides fuel tank vent, electrical, pneumatic, air-conditioning, and preflight conditioning interfaces. Umbilical withdrawal by pneumatic disconnect in conjunction with pneumatic/hydraulic redundant dual cylinder system. Secondary mechanical system. Arm also equipped with line handling device to protect lines during withdrawal. Retract time is 8.4 seconds (max).

⑧ Service Module (in-flight). Provides air-conditioning, vent line, coolant, electrical, and pneumatic interfaces. Umbilical withdrawal by pneumatic/mechanical lanyard system with secondary mechanical system. Retract time is 9.0 seconds (max).

⑨ Command Module Access Arm (preflight). Provides access to spacecraft through environmental chamber. Arm may be retracted or extended from LCC. Retracted 12° park position until T-4 minutes. Extend time is 12 seconds from this position.

MOBILE LAUNCHER SERVICE ARMS

Fig. 23

structure, within the fill, is a one-story concrete building to house the high-pressure gas storage battery. On the pad surface are elevators, staircase, and interface structures to provide service to the ML and the MSS. A ramp, with a five percent grade, provides access from the crawlerway. This is used by the C/T to position the ML/Saturn V and the MSS on the support pedestals. The azimuth alignment building is located on the approach ramp in the crawlerway median strip. A flame trench 58 feet wide by 450 feet long, bisects the pad. This trench opens to grade at the north end. The 700,000-pound, mobile, wedge-type flame deflector is mounted on rails in the trench.

The Pad Terminal Connection Room (PTCR) (Figure 25) provides the terminals for communication and data link transmission connections between the ML or MSS and the launch area facilities and between

LAUNCH PAD A, LC-39

Fig. 24

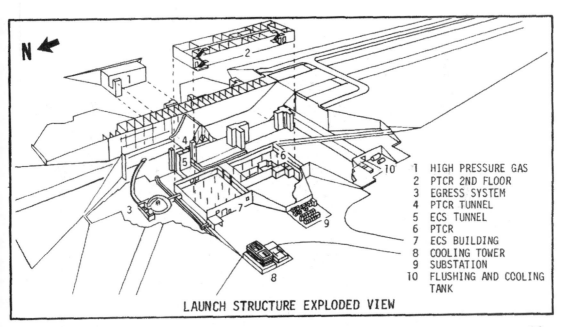

1	HIGH PRESSURE GAS
2	PTCR 2ND FLOOR
3	EGRESS SYSTEM
4	PTCR TUNNEL
5	ECS TUNNEL
6	PTCR
7	ECS BUILDING
8	COOLING TOWER
9	SUBSTATION
10	FLUSHING AND COOLING TANK

LAUNCH STRUCTURE EXPLODED VIEW

Fig. 25

the ML or MSS and the LCC. This facility also accommodates the electronic equipment that simulates functions for checkout of the facilities during the absence of the launcher and vehicle.

The Environmental Control System (ECS) room, located in the pad fill west of the pad structure and north of the PTCR (Figure 25), houses the equipment which furnishes temperature and/or humidity-controlled air or nitrogen for space vehicle cooling at the pad. The ECS room is 96 feet wide by 112 feet long and houses air and nitrogen handling units, liquid chillers, air compressors, a 3000-gallon water/glycol storage

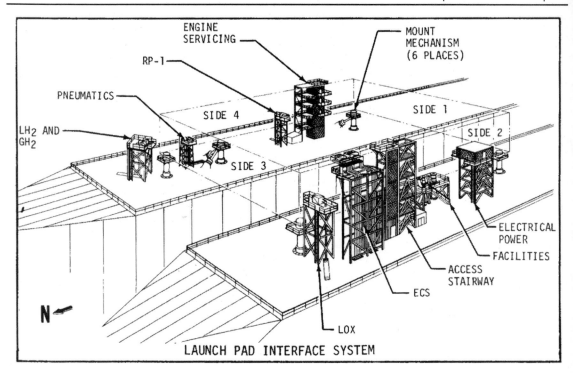

ENGINE
SERVICING

RP-1

PNEUMATICS

MOUNT
MECHANISM
(6 PLACES)

SIDE 4

SIDE 1

LH₂ AND
GH₂

SIDE 2

SIDE 3

ELECTRICAL
POWER

FACILITIES

ACCESS
STAIRWAY

ECS

N

LOX

LAUNCH PAD INTERFACE SYSTEM

Fig. 26

tank, and other auxiliary electrical and mechanical equipment.

The high-pressure gas storage facility at the pad provides the launch vehicle with high-pressure helium and nitrogen. The launch pad interface structure (Figure 26) provides mounting support pedestals for the ML and MSS, an engine access platform, and support structures for fueling, pneumatic, electric power, and environmental control interfaces.

Apollo Emergency ingress/Egress and Escape System

The Apollo emergency ingress/egress and escape system provides access to and from the Command Module (CM) plus an escape route and safe quarters for the astronauts and service personnel in the event of a serious malfunction prior to launch. The system includes the CM access arm, two 600-feet per minute elevators from the 340-foot level to level A of the ML, pad elevator No. 2, personnel carriers located adjacent to the exit of pad elevator No. 2, the escape tube, and the blast room.

The CM access arm provides a passage for the astronauts and service personnel from the spacecraft to the 320-foot level of the tower. Egressing personnel take the high-speed elevators to level A of the ML, proceed through the elevator vestibule and corridor to pad elevator No. 2, move down this elevator to the bottom of the pad, and enter armored personnel carriers which remove them from the pad area.

When the state of the emergency allows no time for retreat by motor vehicle, egressing personnel upon reaching level A of the ML slide down the escape tube into the blast room vestibule, commonly called the "rubber room" (Figure 27).

Entrance to the blast room is gained through blast-proof doors controllable from either side. The blast room floor is mounted on coil springs to reduce outside acceleration forces to between 3 and 5 g's. Twenty people may be accommodated for 24 hours. Communication facilities are provided in the room, including an emergency RF link.

An underground air duct from the vicinity of the blast room to the remote air intake facility permits egress from the pad structure to the pad perimeter. Provision is made to decrease air velocity in the duct to allow personnel movement through the duct.

An alternate emergency egress system employs a slide wire from the vicinity of the 320 foot level of the ML to a 30-foot tower on the ground, approximately 2500 feet west of the launcher. Egressing personnel slide down the wire on individual seat assemblies suspended from the wire. Speeds of approximately 50 miles per hour are attained at the low point approximately 20 feet above ground level. A ferrule at the low point activates a braking mechanism which causes a controlled deceleration of each seat

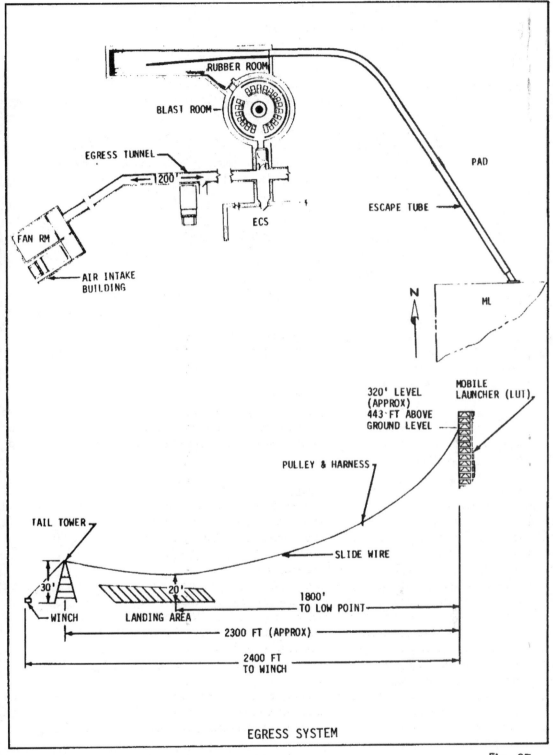

RUBBER ROOM

BLAST ROOM

EGRESS TUNNEL

200'

PAD

ESCAPE TUBE

ECS

FAN RM

AIR INTAKE
BUILDING

N

ML

320' LEVEL
(APPROX)
443 FT ABOVE
GROUND LEVEL

MOBILE
LAUNCHER (LUT)

PULLEY & HARNESS

TAIL TOWER

SLIDE WIRE

30'

20'

1800'
TO LOW POINT

WINCH

LANDING AREA

2300 FT (APPROX)

2400 FT
TO WINCH

EGRESS SYSTEM

Fig. 27

assembly to a safe stop. Up to 11 persons may be accommodated by the system.

Fuel System Facilities

The RP-I facility consists of three 86,000-gallon steel storage tanks, a pump house, a circulating pump, a transfer pump, two filter-separators, an 8-inch stainless steel transfer line, RP-I foam generating building, and necessary valves, piping, and controls. Two RP-I holding ponds (Figure 24), 150 feet by 250 feet, with a water depth of two feet, are located north of the launch pad, one on each side of the north-south axis. The ponds retain spilled RP-I and discharge water to drainage ditches.

The LH facility (Figure 24) consists of one 850,000-gallon spherical storage tank, a vaporizer/heat exchanger which is used to pressurize the storage tank to 65 psi, a vacuum-jacketed, 10-inch, invar transfer line, and a burn pond venting system. Internal tank pressure provides the proper flow of LH from the storage tank to the vehicle without using a transfer pump. Liquid hydrogen boil-off from the storage and ML areas is directed through vent-piping to bubble-capped headers submerged in the burn pond where a hot wire ignition system maintains the burning process.

LOX System Facility

The LOX facility (Figure 24) consists of one 900,000-gallon spherical storage tank, a LOX vaporizer to pressurize the storage tank/ main fill and replenish pumps, a drain basin for venting and dumping of LOX, and two transfer lines.

Azimuth Alignment Building

The azimuth alignment building (Figure 24) houses the auto-collimator theodolite which senses, by a light source, the rotational output of the stable platform in the instrument unit of the launch vehicle. This instrument monitors the critical inertial reference system prior to launch.

Photography Facilities

These facilities support photographic camera and closed circuit television equipment to provide real-time viewing and photographic documentation coverage. There are six camera sites in the launch pad area. These sites cover pre-launch activities and launch operations from six different angles at a radial distance of approximately 1300 feet from the launch vehicle. Each site has four engineering, sequential cameras and one fixed, high-speed metric camera (CZR).

Pad Water System Facilities

The pad water system facilities furnish water to the launch pad area for fire protection, cooling, and quenching. Specifically, the system furnishes water for the industrial water system, flame deflector cooling and quench, ML deck cooling and quench, ML tower fogging and service arm quench, sewage treatment plant, Firex water system, liquid propellant facilities, ML and MSS fire protection, and all fire hydrants in the pad area.

Mobile Service Structure

The MSS (Figure 28) provides access to those portions of the space vehicle which cannot be serviced from the ML while at the launch pad. The MSS is transported to the launch site by the C/T where it is used during launch pad operations. It is removed from the pad a few hours prior to launch and returned to its parking area 7000 feet from the nearest launch pad. The MSS is approximately 402 feet high and weighs 12 million pounds. The tower structure rests on a base 135 feet by 135 feet. At the top, the tower is 87 feet by 113 feet.

The structure contains five work platforms which provide access to the space vehicle. The outboard sections of the platforms open to accept the vehicle and close around it to provide access to the launch vehicle and spacecraft. The lower two platforms are vertically adjustable to serve different parts of the launch vehicle. The upper three platforms are fixed but can be disconnected from the tower and relocated as a unit to serve different vehicle configurations. The second and third platforms from the top are enclosed and provide environmental control for the spacecraft.

The MSS is equipped with the following systems: air conditioning, electrical power, various communication networks, fire protection, compressed air, nitrogen pressurization, hydraulic pressure, potable water, and spacecraft fueling.

Crawler-Transporter

The C/T (Figure 29) is used to transport the ML, including the space vehicle, and the MSS to and from the launch pad.

The C/T is capable of lifting, transporting, and lowering the ML or the MSS, as required, without the aid of auxiliary equipment. The C/T supplies limited electric power to the ML and the MSS during transit.

MOBILE SERVICE STRUCTURE

Fig. 28

CRAWLER TRANSPORTER

Fig. 29

The C/T consists of a rectangular chassis which is supported through a suspension system by four, dual-tread, crawler-trucks. The overall length is 131 feet and the overall width is 114 feet. The unit weighs approximately six million pounds. The C/T is powered by self-contained, diesel-electric generator units. Electric motor-driven pumps provide hydraulic power for steering and suspension control. Air conditioning and ventilation are provided where required.

The C/T can be operated with equal facility in either direction. Control cabs are located at each end. The leading cab, in the direction of travel, has complete control of the vehicle. The rear cab however, has override controls for the rear trucks only.

Maximum C/T speed is 2 mph unloaded, 1 mph with full load on level grade, and 0.5 mph with full load on a five percent grade. It has a 500-foot minimum turning radius and can position the ML or the MSS on the facility support pedestals within ± two inches.

VEHICLE ASSEMBLY AND CHECKOUT

The Launch Vehicle propulsive stages and the IU are, upon arrival at KSC, transported to the VAB by special carriers. The S-IC stage is erected on an ML in one of the checkout bays in the high bay area. Concurrently, the S-11 and S-IVB stages and the IU are delivered to preparation and checkout cells in the low bay area: for inspection, checkout, and pre-erection preparations. All components of the Launch Vehicle, including the Apollo Spacecraft and Launch Escape System, are then assembled vertically on the ML in the high bay area.

Following assembly, the space vehicle is connected to the LCC via a high-speed data link for integrated checkout and a simulated flight test. When checkout is completed, the C/T picks up the ML with the assembled space vehicle and moves it to the launch site via the crawlerway.

At the launch site, the ML is emplaced and connected to system interfaces for final vehicle checkout and launch monitoring. The MSS is transported from its parking area by the C/T and positioned on the side of the vehicle opposite the ML. A flame deflector is moved on its track to its position beneath the blast opening of the ML to deflect the blast from the S-1C stage engines. During the pre-launch checkout, the final system checks are completed, the MSS is removed to the parking area, propellants are loaded, various items of support equipment are removed from the ML, and the vehicle is readied for launch. After vehicle launch, the C/T transports the ML to the parking area near the VAB for refurbishment.

MISSION MONITORING, SUPPORT, AND CONTROL GENERAL

Mission execution involves the following functions: pre-launch checkout and launch operations; tracking the space vehicle to determine its present and future positions; securing information on the status of the flight crew and space vehicle systems (via telemetry); evaluation of telemetry information commanding the space vehicle by transmitting real-time and update commands to the onboard computer; voice communication between flight and ground crews; and recovery operations.

These functions require the use of a facility to assemble and launch the space vehicle (see Launch Complex); a central flight control facility; a network of remote stations located strategically around the world; a method of rapidly transmitting and receiving information between the space vehicle and the central flight control facility; a real-time data display system in which the data is made available and presented in usable form at essentially the same time that the data event occurred; and ships/aircraft to recover the spacecraft on return to earth.

The Flight crew and the following organizations and facilities participate in mission control operations:

1. Mission Control Center (MCC), Manned Spacecraft Center (MSC), Houston, Texas. The MCC contains the communication, computer, display, and command systems to enable the flight controllers to effectively monitor and control the space vehicle.

2. Kennedy Space Center (KSC), Cape Kennedy, Florida. The space vehicle is launched From KSC and controlled from the Launch Control Center (LCC), as described previously. Pre-launch, launch, and powered flight data are collected at the Central Instrumentation Facility (CIF) at KSC from the launch pads, CIF receivers, Merritt Island Launch Area (MILA), and the down-range Air Force Eastern Test Range (AFETR) stations. This data is transmitted to MCC via the Apollo Launch Data System (ALDS). Also located at KSC (ETR) is the Impact Predictor (IP), for range safety purposes.

3. Goddard Space Flight Center (GSFC), Greenbelt, Maryland. GSFC manages and operates the Manned Space Flight Network (MSFN) and the NASA communications (NASCOM) networks. During flight, the MSFN is under operational control of the MCC.

4. George C. Marshall Space Flight Center (MSFC), Huntsville, Alabama. MSFC, by means of the Launch information Exchange Facility (LIEF) and the Huntsville Operations Support Center (HOSC), provides launch vehicle systems real-time support to KSC and MCC for pre-flight, launch, and flight operations.

A block diagram of the basic flight control interfaces is shown in Figure 30.

VEHICLE FLIGHT CONTROL CAPABILITY

Flight operations are controlled from the MCC. The MCC has two flight control rooms. Each control room, called a Mission Operations Control Room (MOCR), is used independently of the other and is capable of controlling individual Staff Support Rooms (SSR's) located adjacent to the MOCR. The SSR's are manned by flight control specialists who provide detailed support to the MOCR. Figure 31 outlines the

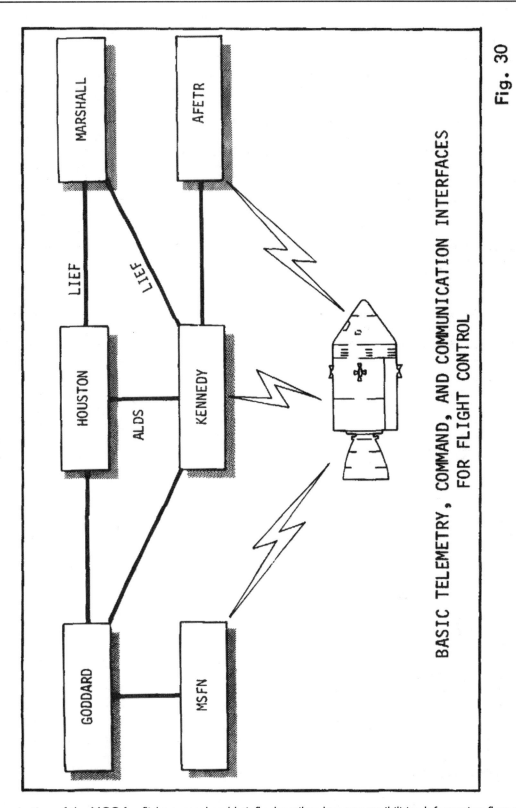

BASIC TELEMETRY, COMMAND, AND COMMUNICATION INTERFACES FOR FLIGHT CONTROL

Fig. 30

organization of the MCC for flight control and briefly describes key responsibilities. Information flow within the MOCR is shown in Figure 32.

The consoles within the MOCR and SSR's permit the necessary interface between the flight controllers and the spacecraft. The displays and controls on these consoles and other group displays provide the capability to monitor and evaluate data concerning the mission and, based on these evaluations, to

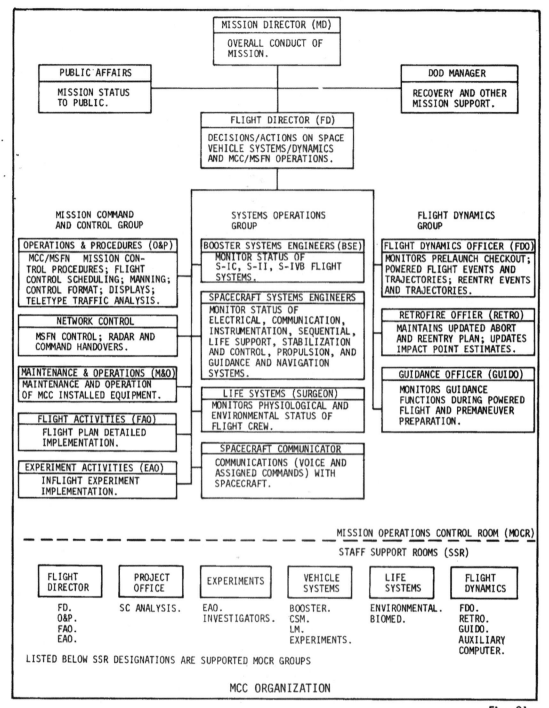

Fig. 31

recommend or take appropriate action on matters concerning the flight crew and spacecraft.

Problems concerning crew safety and mission success are identified to flight control personnel in the following ways:

Flight crew observations; Flight controller real-time observations; Review of telemetry data received from tape recorder playback; Trend analysis of actual and predicted values; Review of collected data by systems specialists; Correlation and comparison with previous mission data; Analysis of recorded data from launch complex testing.

The facilities at the MCC include an input/output processor designated as the Command,

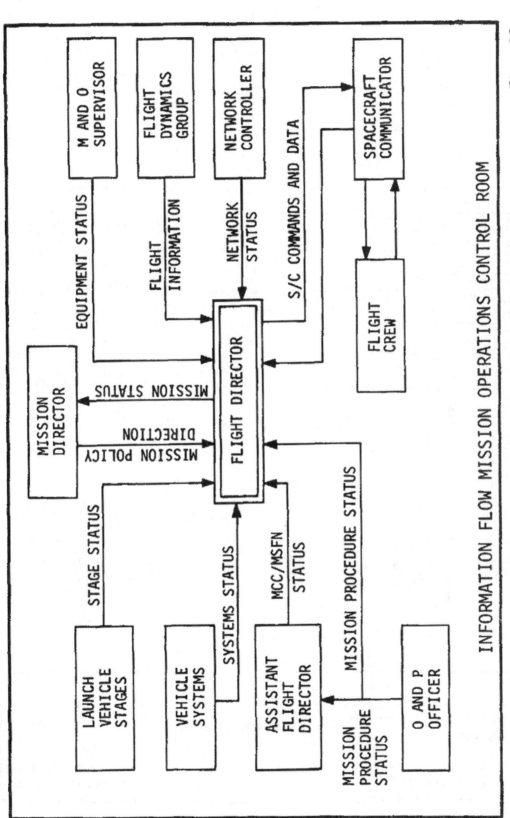

Fig. 32

INFORMATION FLOW MISSION OPERATIONS CONTROL ROOM

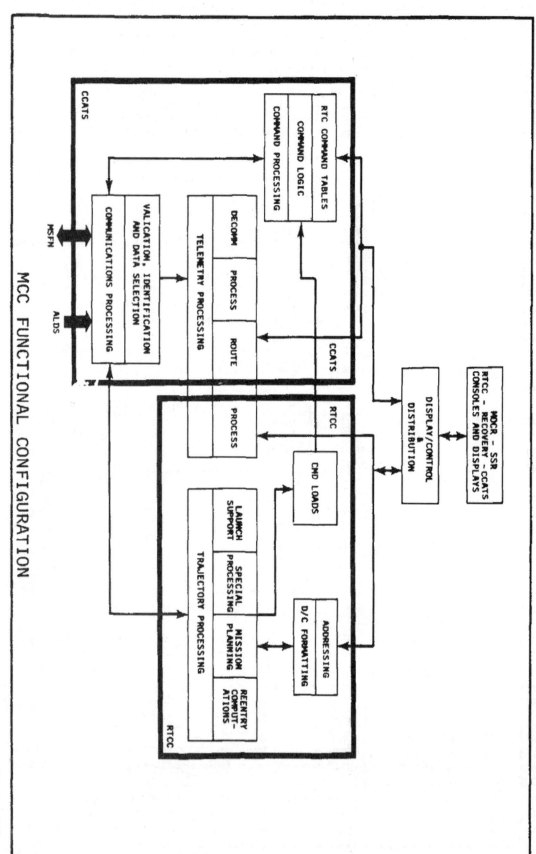

MCC FUNCTIONAL CONFIGURATION

Fig. 33

Communications, and Telemetry System (CCATS) and a computational facility, the Real-Time Computer Complex (RTCC). Figure 33 shows the MCC functional configuration.

The CCATS consists of three Univac 494 general purpose computers. Two of the computers are configured so that either may handle all of the input/output communications for two complete missions. One of the computers acts as a dynamic standby. The third computer is used for non mission activities.

The RTCC is a group of five IBM 360, large scale general purpose computers. Any of the five computers may be designated as the Mission Operations Computer (MOC). The MOC performs all the required computations and display formatting for a mission. One of the remaining computers will be a dynamic standby. Another pair of computers may be used for a second mission or simulation.

Space Vehicle Tracking

From lift-off of the launch vehicle to insertion into orbit, accurate position data are required to allow the Impact Predictor (IP) to function effectively as a Range Safety device, and the RTCC to compute a trajectory and an orbit. These computations are required by the flight controllers to evaluate the trajectory, the orbit, and/or any abnormal situations to ensure safe recovery of the astronauts. The launch tracking date are transmitted from the AFETR site to the IP and thence to the RTCC via high speed data communications circuits. The IP also generates spacecraft inertial positions and inertial rates of motion in real-time.

During boost the trajectory is calculated and displayed on consoles and plot boards in the MOCR and SSR's. Also displayed are telemetry data concerning status of launch vehicle and spacecraft systems. if the space vehicle deviates excessively from the nominal flight path, or if any critical vehicle condition exceeds tolerance limits, or if the safety of the astronauts or range personnel is endangered, a decision is made to abort the mission.

During the orbit phase of a mission, all stations that are actively tracking the spacecraft will transmit the tracking data through GSFC to the RTCC by teletype. If a thrusting maneuver is performed by the spacecraft, high-speed tracking data is also transmitted.

Command System

The Apollo ground command systems have been designed to work closely with the telemetry and trajectory systems to provide flight controllers with a method of "closed loop" command. The astronauts and flight controllers act as links in this operation.

To prevent spurious commands from reaching the space vehicle, switches on the Command Module console block uplink data from the onboard computers. At the appropriate times, the flight crew will move the switches from the "BLOCK" to the "ACCEPT" positions and thus permit the flow of uplink data. With a few exceptions, commands to the space vehicle fall into two categories: real-time commands, and command loads (also called computer loads, computer update, loads, or update).

Real-time commands are used to control space vehicle systems or subsystems from the ground. The execution of a real-time command results in immediate reaction by the affected system. Real-time commands are stored prior to the mission in the Command Data Processor (CDP) at the applicable command site. The CDP, a Univac 642B, general purpose digital computer, is programmed to format, encode, and output commands when a request for uplink is generated.

Command loads are generated by the real-time computer complex on request of flight controllers. Command loads are based on the latest available telemetry and/or trajectory data.

Flight controllers typically required to generate a command load include the Booster Systems Engineer (BSE), the Flight Dynamics Officer (FDO), the Guidance Officer (GUIDO), and the Retrofire Officer (RETRO).

Display and Control System

The MCC is equipped with facilities which provide for the input of data from the MSFN and KSC over a combination of high-speed data, low-speed data, wide-band data, teletype, and television channels. These data are computer processed for display to the flight controllers.

Several methods of displaying data are used including television (projection TV, group displays, closed circuit TV, and TV monitors), console digital readouts, and event lights. The display and control system interfaces with the RTCC and includes computer request, encoder multiplexer, plotting display, slide file, digital-to-TV converter, and telemetry event driver equipments.

A control system is provided for flight controllers to exercise their respective functions for mission control and technical management. This system is comprised of different groups of consoles with television monitors, request keyboards, communications equipment, and assorted modules added as required to provide each operational position in the MOCR with the control and display capabilities required for the particular mission.

CONTINGENCY PLANNING AND EXECUTION

Planning for a mission begins with the receipt of mission requirements and objectives. The planning activity results in specific plans for pre-launch and launch operations, preflight training and simulation, flight control procedures, flight crew activities,

MSFN and MCC support, recovery operations, data acquisition and flow, and other mission-related operations. Numerous simulations are planned and performed to test procedures and train flight control and flight crew teams in normal and contingency operations.

MCC Role in Aborts

After launch and from the time the space vehicle clears the ML, the detection of slowly deteriorating conditions which could result in an abort is the prime responsibility of MCC; prior to this time, it is the prime responsibility of LCC. In the event such conditions are discovered, MCC requests abort of the mission or, circumstances permitting, sends corrective commands to the vehicle or requests corrective flight crew actions.

In the event of a non catastrophic contingency, MCC recommends alternate flight procedures, and mission events are rescheduled to derive maximum benefit from the modified mission.

VEHICLE FLIGHT CONTROL PARAMETERS

In order to perform flight control monitoring functions, essential data must be collected, transmitted, processed, displayed, and evaluated to determine the space vehicle's capability to start or continue the mission.

Parameters Monitored by LCC

The launch vehicle checkout and pre-launch operations monitored by the Launch Control Center (LCC) determine the state of readiness of the launch vehicle, ground support, telemetry, range safety, and other operational support systems. During the final countdown, hundreds of parameters are monitored to ascertain vehicle, system, and component performance capabilities. Among these parameters are the "redlines." The redline values must be within the predetermined limits or the countdown will be halted. In addition to the redlines, there are a number of operational support elements such as ALDS, range instrumentation, ground tracking and telemetry stations, and ground support facilities which must be operational at specified times in the countdown.

Parameters Monitored by Booster Systems Group

The Booster Systems Group (BSG) monitors launch vehicle systems (S-IC, S-II, S-IVB, and IU) and advises the flight director and flight crew of any system anomalies. It is responsible for confirming in-flight power, stage ignition, holddown release, all engines go, engine cut-offs, etc. BSG also monitors attitude control, stage separations, and digital commanding of LV systems.

Parameters Monitored by Flight Dynamics Group

The Flight Dynamics Group monitors and evaluates the powered flight trajectory and makes the abort decisions based on trajectory violations. It is responsible for abort planning, entry time and orbital maneuver determinations, rendezvous planning, inertial alignment correlation, landing point prediction, and digital commanding of the guidance systems.

The MOCR positions of the Flight Dynamics Group include the Flight Dynamics Officer (FDO), the Guidance Officer (GUIDO), and the Retrofire Officer (RETRO). The MOCR positions are given detailed, specialized support by the Flight Dynamics SSR.

The surveillance parameters measured by the ground tracking stations and transmitted to the MCC are computer processed into plot board and digital displays. The Flight Dynamics Group compares the actual data with pre-mission, calculated, nominal data and is able to determine mission status.

Parameters Monitored by Spacecraft Systems Group

The Spacecraft Systems Group monitors and evaluates the performance of spacecraft electrical, optical, mechanical, and life support systems; maintains and analyzes consumables status; prepares the mission log; coordinates telemetry playback; determines spacecraft weight and center of gravity; and executes digital commanding of spacecraft systems.

The MOCR positions of this group include the Command and Service Module Electrical, Environmental, and Communications Engineer (CSM EECOM), the CSM Guidance, Navigation, and Control Engineer (CSM GNC), the Lunar Module Electrical, Environmental, and Communications Engineer (LM EECOM), and the LM Guidance, Navigation, and Control Engineer (LM GNC). These positions are backed up with detailed support from the Vehicle Systems SSR.

Parameters Monitored by Life Systems Group

The Life Systems Group is responsible for the well-being of the flight crew. The group is headed by the Flight Surgeon in the MOCR. Aeromedical and environmental control specialists in the Life Systems SSR provide detailed support to the Flight Surgeon. The group monitors the Flight crew health status and environmental/ biomedical parameters.

MANNED SPACE FLIGHT NETWORK

The Manned Space Flight Network (MSFN) is a global network of ground stations, ships, and aircraft designed to support manned and unmanned space flights. The network provides tracking, telemetry, voice and teletype communications, command, recording, and television capabilities. The network is specifically configured to meet the requirements of each mission.

MSFN stations are categorized as lunar support stations (deep-space tracking in excess of 15,000 miles), near-space support stations with Unified S-Band (USB) equipment, and near-space support stations without USB equipment. Figure 34 shows the geographical location of each station.

MSFN stations include facilities operated by NASA, the United States Department of Defense (DOD), and the Australian Department of Supply (DOS).

The DOD facilities include the Eastern Test Range (ETR), Western Test Range (WTR), White Sands Missile Range (WSMR), Range Instrumentation Ships (RIS), and Apollo Range Instrumentation Aircraft (A/RIA).

NASA COMMUNICATION NETWORK

The NASA Communications (NASCOM) network (Figure 35) is a point-to-point communications systems connecting the MSFN stations to the MCC. NASCOM is managed by the Goddard Space Flight Center, where the primary communications switching center is located. Three smaller NASCOM switching centers are located at London, Honolulu, and Canberra. Patrick AFB, Florida and Wheeler AFB, Hawaii serve as switching centers for the DOD eastern and western test ranges, respectively. The MSFN stations throughout the world are interconnected by land line, undersea cable, radio and communications satellite circuits. These circuits carry teletype, voice, and data in real-time support of the missions.

Each MSFN USB land station has a minimum of five voice/data circuits and two teletype circuits. The Apollo insertion and injection ships have a similar capability through the communications satellites.

APOLLO LAUNCH DATA SYSTEM (ALDS)

The Apollo Launch Data System (ALDS) between KSC and MSC is controlled by MSC and is not routed through GSFC. The ALDS consists of wide-band telemetry, voice coordination circuits, and a high-speed circuit for the Countdown and Status Transmission System (CASTS). In addition, other circuits are provided for launch coordination, tracking data, simulations, public information, television, and recovery.

MSFC SUPPORT FOR LAUNCH AND FLIGHT OPERATIONS

The Marshall Space Flight Center (MSFC), by means of the Launch Information Exchange Facility (LIEF) and the Huntsville Operations Support (HOSC) provides real-time support of launch vehicle pre-launch, launch, and flight operations. MSFC also provides support, via LIEF, for post-flight data delivery and evaluation.

In-depth real-time support is provided for pre-launch, launch, and flight operations from HOSC consoles manned by engineers who perform detailed system data monitoring and analysis.

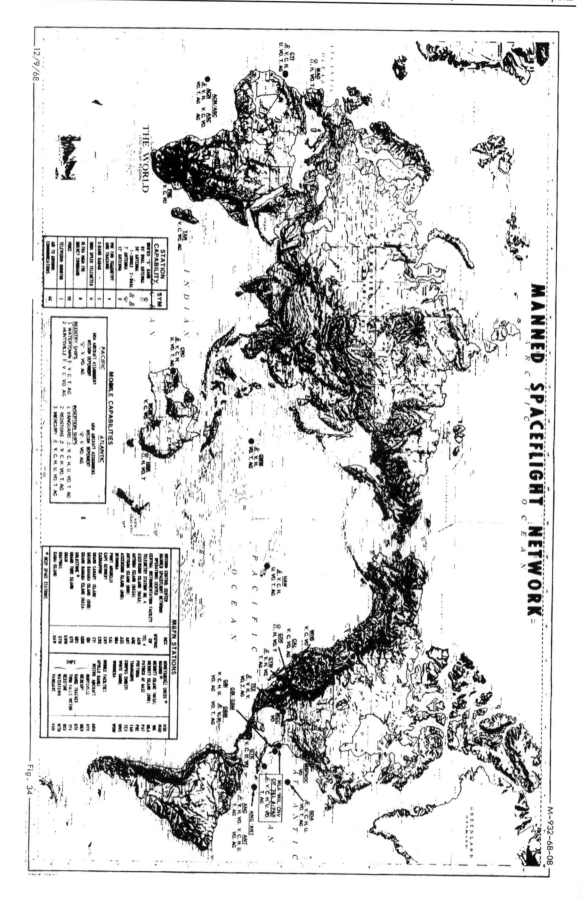

Fig. 34

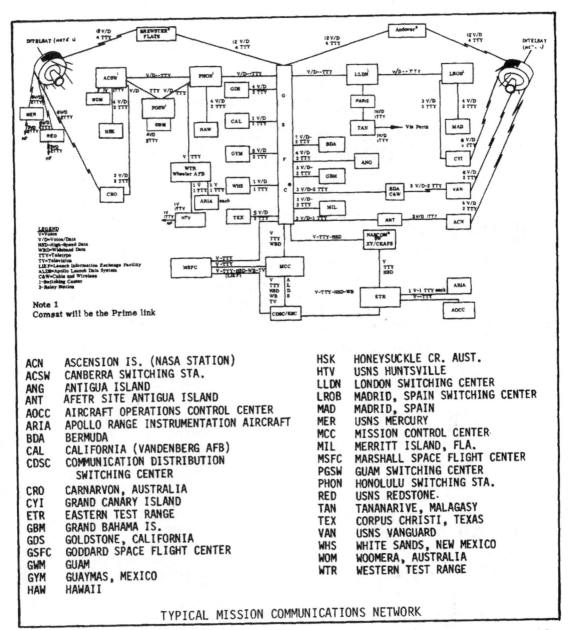

LEGEND
V=Voice
V/D=Voice/Data
HSD=High-Speed Data
WBD=Wideband Data
TTY=Teletype
TV=Television
LIEF=Launch Information Exchange Facility
ALDS=Apollo Launch Data System
C&W=Cable and Wireless
1=Switching Center
2=Relay Station

Note 1
Comsat will be the Prime link

ACN	ASCENSION IS. (NASA STATION)	HSK	HONEYSUCKLE CR. AUST.
ACSW	CANBERRA SWITCHING STA.	HTV	USNS HUNTSVILLE
ANG	ANTIGUA ISLAND	LLDN	LONDON SWITCHING CENTER
ANT	AFETR SITE ANTIGUA ISLAND	LROB	MADRID, SPAIN SWITCHING CENTER
AOCC	AIRCRAFT OPERATIONS CONTROL CENTER	MAD	MADRID, SPAIN
ARIA	APOLLO RANGE INSTRUMENTATION AIRCRAFT	MER	USNS MERCURY
BDA	BERMUDA	MCC	MISSION CONTROL CENTER
CAL	CALIFORNIA (VANDENBERG AFB)	MIL	MERRITT ISLAND, FLA.
CDSC	COMMUNICATION DISTRIBUTION	MSFC	MARSHALL SPACE FLIGHT CENTER
	SWITCHING CENTER	PGSW	GUAM SWITCHING CENTER
CRO	CARNARVON, AUSTRALIA	PHON	HONOLULU SWITCHING STA.
CYI	GRAND CANARY ISLAND	RED	USNS REDSTONE.
ETR	EASTERN TEST RANGE	TAN	TANANARIVE, MALAGASY
GBM	GRAND BAHAMA IS.	TEX	CORPUS CHRISTI, TEXAS
GDS	GOLDSTONE, CALIFORNIA	VAN	USNS VANGUARD
GSFC	GODDARD SPACE FLIGHT CENTER	WHS	WHITE SANDS, NEW MEXICO
GWM	GUAM	WOM	WOOMERA, AUSTRALIA
GYM	GUAYMAS, MEXICO	WTR	WESTERN TEST RANGE
HAW	HAWAII		

TYPICAL MISSION COMMUNICATIONS NETWORK

Fig. 35

Pre-launch flight wind monitoring analysis and trajectory simulations are jointly performed by MSFC and MSC personnel located at MSFC during the terminal countdown. Beginning at T-24 hours, actual wind data is transmitted periodically from KSC to the HOSC. These measurements are used by the MSFC/MSC wind monitoring team in vehicle flight digital simulations to verify the capability of the vehicle with these winds.

In the event of marginal wind conditions, contingency data are provided MSFC in real-time via the Central Instrumentation Facility (CIF). DATA-CORE and trajectory simulations are performed on-line to expedite reporting to KSC.

During the pre-launch period, primary support is directed to KSC. At lift-off primary support transfers from KSC to the MCC. The HOSC engineering consoles provide support as required to the Booster Systems Group for S-IVB/IU orbital operations by monitoring detailed instrumentation for the evaluation of system in-flight and dynamic trends, assisting in the detection and isolation of vehicle malfunctions and providing advisory contact with vehicle design specialists.

ABBREVIATIONS AND ACRONYMS

ac	Alternating Current
AFB	Air Force Base
AFETR	Air Force Eastern Test Range
ALDS	Apollo Launch Data System
AM	Amplitude Modulation
APS	Auxiliary Propulsion System
ARIA	Apollo Range Instrumentation Aircraft
AS	Apollo Saturn
ASI	Augmented Spark Igniter
BPC	Boost Protective Cover
BSE	Booster Systems Engineer
CASTS	Countdown and Status Transmission System
CCATS	Communications, Command, and Telemetry System
CCS	Command Communications System
CDP	Command Data Processor (MSFN Site)
CIF	Central Instrumentation Facility
CM	Command Module
COAS	Crewman Optical Alignment Sight
CSM	Command Service Module
C/T	Crawler/Transporter
CWG	Constant-Wear Garment
DATA-CORE	CIF Telemetry Conversion System
dc	Direct Current
DOD	Department of Defense
DOS	Department of Supply (Australia)
ECS	Environmental Control System
EDS	Emergency Detection System
ELS	Earth Landing System
EMS	Entry Monitor System
EMU	Extravehicular Mobility Unit
EFS	Electrical Power System
ETR	Eastern Test Range
EV	Extravehicular
EVA	Extravehicular Activity
KC	Flight Control Computer (IU, analog)
FDAI	Flight Director Attitude Indicator
FDO	Flight Dynamics Officer
g	Force of gravity (local)
GDC	Gyro Display Coupler
GH2	Gaseous Hydrogen
GN2	Gaseous Nitrogen
GNCS	Guidance, Navigation, and Control System
GOX	Gaseous Oxygen
GSE	Ground Support Equipment
GUIDO	Guidance Officer
GSFC	Goddard Space Flight Center
H2	Hydrogen
HF	High Frequency
HOSC	Huntsville Operations Support Center
ICG	Inflight Coverall Garment
IMU	Inertial Measurement Unit
IP	Impact Predictor (at KSC)
IU	Instrument Unit
KSC	Kennedy Space Center
LC	Launch Complex
LCC	Launch Control Center
LCG	Liquid-Cooling Garment
LEA	Launch Escape Assembly
LEB	Lower Equipment Bay
LES	Launch Escape System
LET	Launch Escape Tower
LH	Liquid Hydrogen
LIEF	Launch information Exchange Facility
LM	Lunar Module
LN2	Liquid Nitrogen
LOX	Liquid Oxygen
LV	Launch Vehicle

LVDA	Launch Vehicle Data Adapter
LVDC	Launch Vehicle Digital Computer
MCC	Mission Control Center
MILA	Merritt Island Launch Area
ML	Mobile Launcher
MMH	Monomethyl Hydrazine
MOC	Mission Operations Computer
MOCR	Mission Operations Control Room
MSC	Manned Spacecraft Center
MSFC	Marshall Space Flight Center
MSFN	Manned Space Flight Network
MSS	Mobile Service Structure
NASCOM	NASA Communications Network
N2O4	Nitrogen Tetroxide
NPSH	Net Positive Suction Head
O2	Oxygen
OMR	Operations Management Room
OSR	Operations Support Room
PDS	Propellant Dispersion System
PGA	Pressure Garment Assembly
PLSS	Portable Life Support System
PTC	Pad Terminal Connection Room
RPU	Propellant Utilization
RCS	Reaction Control System
RETRO	Direction Opposite to Velocity Vector
RF	Radio Frequency
RIS	Range Instrumentation Ship
RP-1	Rocket Propellant (refined kerosene)
RTCC	Real Time Computer Complex
SC	Apollo Spacecraft
SCS	Stabilization and Control System
SECS	Sequential Events Control System
SLA	Spacecraft Lunar Module Adapter
SM	Service Module
SPS	Service Propulsion System
SSR	Staff Support Room
SV	Space Vehicle
TCS	Thermal Conditioning System
TSM	Tail Service Mast
TV	Television
USB	Unified S-band
UHF	Ultra-High Frequency
VAB	Vehicle Assembly Building
VHF	Very High Frequency
WSMR	White Sands Missile Range
WTR	Western Test Range

NASA Mission Objectives
For Apollo 8

MEMORANDUM

To:A/Acting Administrator
From : MA/Apollo Program Director

10 February 1969

Subject:Apollo 8 Mission (AS-503) Post Launch Report # 1

    The Apollo 8 mission was successfully launched from the Kennedy Space Center on Saturday, 21 December 1968 and completed as planned with recovery of the spacecraft and crew in the Pacific recovery area on Friday, 27 December 1968. Initial evaluation of the flight based upon quick-look data and crew debriefing indicates that all mission objectives were attained. Further detailed analysis of all data is continuing and appropriate refined results of the mission will be reported in Manned Space Flight Center technical reports based on mission performance as described in this report, I am recommending that the Apollo 8 mission be adjudged as having achieved agency preset primary objectives and considered a success.

Sam Phillips
Lt. General, USAF
Apollo Program Director

Approval
George E. Mueller
Associate Administrator for Manned Space Flight

NASA MISSION OBJECTIVES FOR APOLLO 8

PRIMARY OBJECTIVES

Demonstrate crew/space vehicle/mission support facilities performance during a manned Saturn V mission with CSM.

Demonstrate performance of nominal and selected backup Lunar Orbit Rendezvous (LOR) mission activities, including:
- Translunar injection
- CSM navigation, communications, and midcourse corrections
- CSM consumables assessment and passive thermal control

Sam Phillips
Lt. General, USAF
Apollo Program Director
Date: 12 Dec 1968

George E. Mueller
Associate Administrator for Manned Space Flight

Date: 13 Dec 1968

RESULTS OF APOLLO 8 MISSION

Based upon review of the assessed performance of Apollo 8, launched 21 December 1968 and completed 27 December 1968, this mission is adjudged a success in accordance with the pre-set objectives stated above.

Sam C. Phillips
Lt. General. USAF
Apollo Program Director
Date: 10 Feb 1969

George E. Mueller
Associate Administrator for Manned Space Flight

Date: 10 Feb 1969

GENERAL

The Apollo 8 (AS-503) mission was launched from the Kennedy Space Center on Saturday, 21 December 1968, at 7:51 a.m. EST. Initial data review indicates that all mission objectives were successfully accomplished, as well as four detailed test objectives that were not originally planned. Only minor anomalies and discrepancies occurred and these will be discussed in succeeding sections.

COUNTDOWN

The terminal countdown for Apollo 8 began at 21:51 EST on 20 December. Built-in holds of 6 hours at T-9 hours and 1 hour at T-3 hours 30 minutes were initiated as planned and no additional holds were required. Launch occurred on time at the opening of the launch window at 07:51 EST, 21 December.

MISSION SUMMARY

The Apollo 8 space vehicle was launched from Launch Complex 39, Pad A, at the Kennedy Space Center at 07:51 EST on 21 December 1968, after a satisfactory countdown. Following a nominal boost phase, the spacecraft and S-IVB combination was inserted into a parking orbit of 98.0 by 103.3 nautical miles (NM). After a post-insertion checkout of spacecraft systems, the Translunar Injection (TLI) maneuver was initiated at 02:50:37 Ground Elapsed Time (GET) by reigniting the S-IVB engine and burning for 5 minutes 9 seconds.

The spacecraft separated from the S-IVB at 03:20:59 GET, and performed separation maneuvers using the Service Module Reaction Control System (SM RCS). The first midcourse correction of 24.8 feet per second (fps) was conducted at 11:00:00 GET. The translunar coast phase was devoted to navigation sightings, two television transmissions, and various systems checks. A second midcourse correction of 1.4 fps was conducted at 60:59:56 GET.

The 246.9-second duration Service Propulsion System (SPS) lunar orbit insertion maneuver was performed at 69:08:20 GET, and the initial lunar orbit was 168.5 by 59.9 NM. An SPS maneuver to circularize the orbit was conducted at 73:35:07 GET and resulted in a lunar orbit of 59.7 by 60.7 NM. A total of 10 revolutions was completed during the 20 hours 11 minutes spent in lunar orbit.

The lunar orbit coast phase between maneuvers involved numerous landing-site/landmark sightings, photography, two television transmissions, and preparation for Transearth Injection (TEI). The TEI aneuver, 203.7 seconds in duration, was conducted at 89:19:17 GET, using the SPS.

During both translunar and transearth coast phases, passive thermal control roll maneuvers of about ne revolution per hour were effected, when possible, to maintain system temperatures within nominal nits. The transearth coast period involved a number of star/horizon navigation sightings, using both the arth and moon horizons, and two additional television transmissions. The only transearth midcourse orrection was a 4.8 fps maneuver made at 103:59:54 GET.

Command Module/Service Module separation was at 146:28:48 GET, and the spacecraft reached the ntry interface (400,000 feet) at 146:46:13 GET. Entry conditions were a velocity of 36,221 fps (36,219 fps lanned) and a flight path angle of -6.5 degrees, Following normal deployment of all parachutes, the spacecraft nded in the Pacific Ocean at 08°07.5'N latitude and 165°01.2'W longitude, less than one nautical mile from ne predicted splashdown point. The total flight duration was 147 hours 00 minutes 42 seconds, and the pacecraft and crew were recovered by the USS YORKTOWN after landing.

Almost without exception, spacecraft systems operated as intended. All temperatures varied within cceptable limits and essentially exhibited predicted behavior. Consumables usage was always maintained at safe level. Communications quality was exceptionally good, and live television was transmitted on six ccasions. The crew satisfactorily performed all flight plan functions and achieved all photographic objectives. ummaries of mission events, orbits, and maneuvers are presented in Tables I, II, and III.

TABLE I
SUMMARY OF MISSION EVENTS

EVENT	TIME, HR:MIN:SEC PLANNED	GET ACTUAL
Launch Phase Range Zero (12:51:00 GMT)	00:00:00	00:00:00
Lift-off	00:00:01	00:00:01
Max Q (Maximum Dynamic Pressure)	00:01:16	00:01:19
S-IC center engine cutoff	00:02:06	00:02:06
S-IC outboard engine cutoff	00:02:31	00:02:34
S-IC/S-II separation	00:02:32	00:02:34
S-II engine ignition	00:02:33	00:02:35
Interstage jettison (2nd plane separation)	00:03:02	00:03:04
Launch escape tower jettison	00:03:08	00:03:09
S-II engine cutoff	00:08:41	00:08:44
S-II/S-IVB separation	00:08:42	00:08:45
S-IVB engine ignition	00:08:42	00:08:45
S-IVB engine cutoff	00:11:24	00:11:25
Insertion into earth parking orbit	00:11:34	00:11:35
Orbital Phase Translunar injection ignition	02:50:37	02:50:37
Translunar injection cutoff	02:55:53	02:55:56
S-IVB/Command Module separation	03:20:52	03:20:59
Separation maneuver 1	03:40:00	03:40:01
Separation maneuver 2	Not planned	04:45:01
Start S-IVB propellant dump	05:07:53	05:07:56
End S-IVB Propellant dump	05:12:53	05:12:56
Midcourse correction 1 ignition	11:00:00	11:00:00
Midcourse correction 1 cutoff	11:00:02	11:00:02
Midcourse correction 2 ignition	61:08:51	60:59:56
Midcourse correction 2 cutoff	61:09:03	61:00:08
Lunar orbit insertion 1 ignition	69:08:51	69:08:20
Lunar orbit insertion 1 cutoff	69:11:36	69:12:27
Lunar orbit insertion 2 ignition	73:32:17	73:35:07
Lunar orbit insertion 2 cutoff	73:32:27	73:35:16
Transearth injection ignition	89:16:10	89:19:17
Transearth injection cutoff	89:19:28	89:22:40
Midcourse correction 3 ignition	104:16:10	103:59:54
Midcourse correction 3 cutoff	104:16:24	104:00:08
CM/SM separation	146:31:13	146:28:48
Entry interface (400,000 ft)	146:46:13	146:46:13
Begin blackout	146:46:38	146:46:37
End blackout	146:51:44	146:51:42
Drogue parachute deployment	146:54:26	146:54:48
Main parachute deployment	146:55:19	146:55:39
Landing	146:59:49	147:00:42

TABLE II
ORBIT SUMMARY

PARAMETER	EARTH ORBIT INSERTION INSERTION 1		LUNAR ORBIT INSERTION 2		LUNAR ORBIT	
	Planned	Actual	Planned	Actual	Planned	Actual
Apoapsis (NM)	103.3	103.3	170	168.5	60	60.7
Periapsis (NM)	99.4	98.0	60	59.9	60	59.7
Period (Min)	88.19	88.19	128.7	128.7	119.0	119.0
Inclination (Deg)	32.5	32.5	12.29	12.29	12.29	12.29

TABLE III
MANEUVER SUMMARY

MANEUVER	SYSTEM	TIME HR MIN.SEC Planned (a)	Actual	DURATION SEC Planned (a)	Actual	VELOCITY CHANGE FPS Planned (a)	Actual
Translunar Injection	S-IVB	02:50:37	02:50:37	316	318	35,548(b)	35,532(b)
Separation 1	RCS	03:40:00	03:40:01	(c)	(c)	1.5	1.1
Separation 2	RCS	Not planned	04:45:01	(c)	(c)	9.0	7.7
Midcourse Correction 1	SPS/RCS	11:00:00	11:00:00	2.4	2.4	24	24.8
Midcourse Correction 2	RCS	61:08:51	60:59:56	12.0	11.8	2.0	1.4
Lunar Orbit Insertion 1	SPS	69:08:51	69:08:20	240	246.9	2,992	2,997
Lunar Orbit Insertion 2	SPS	73:32:17	73:35:07	9.5	9.6	138.5	134.8
Transearth Injection	SPS	89:16:10	89:19:17	198	203.7	3,531	3,519
Midcourse Correction 3	RCS	104:16:10	103:59:54 14	13.7		5.0	4.8

Notes: (a) Planned values are those planned prelaunch except for midcourse correction durations and velocity changes. These "planned" values represent real-time updates. (b) These values are velocity at TLI cutoff. c) Burn times for the RCS maneuvers shown are not known at this time.

MEMORANDUM

To: A/Acting Administrator
From: MA/Apollo Program Director
10 February 1969

Subject: Apollo 8 Mission (AS-503) Post Launch Report # 1

The Apollo 8 mission was successfully launched from the Kennedy Space Center on Saturday, 21 December 1968 and completed as planned with recovery of the spacecraft and crew in the Pacific recovery area an Friday, 27 December 1968. Initial evaluation of the flight based upon quick-look data and crew debriefing indicates that all mission objectives were attained. Further detailed analysis of all data is continuing and appropriate refined results of the mission will be reported in Manned Space Flight Center technical reports.

Based on mission performance as described in this report, I am recommending that the Apollo 8 mission be adjudged as having achieved agency preset primary objectives and considered a success.

Sam C. Phillips
Lt. General, USAF
Apollo Program Director

George E. Mueller
Associate Administrator for
Manned Space Flight

LAUNCH VEHICLE

Early engineering evaluation of the SA-503 Saturn V Launch Vehicle indicates that all test and mission objectives were satisfactorily met. All systems and subsystems appear to have performed near nominal. Further evaluation is being carried out to determine detail performance.

Performance of the S-IC POGO suppression system was nominal and indications are that no POGO existed.

The S-IC outboard engines were cut off 2.4 seconds later than predicted by a fuel level cutoff signal, ntly due to higher than predicted fuel density.

Overall propulsion system performance was near nominal, but near the end of S-II stage burn, oscillations of approximately 18 hertz became evident in engine number 5 parameters, beginning at approximately 451 seconds elapsed time. The oscillations became more pronounced at 480 seconds when a small drop in engine number 5 performance occurred, and they damped out shortly before engine cutoff. There also appears to have been some low amplitude, 18-hertz oscillations in the outboard engine pressures. Accelerometers showed 9 to 11-hertz structural oscillations of small amplitudes in both the longitudinal and lateral directions near the same time period.

Two S-II stage temperature bridge power supplies operated intermittently for about 30 seconds near "Max Q." and another similar S-II power supply operated intermittently from 443 to 470 seconds GET.

The modified augmented spark igniter (ASI) lines on both the S-II and S-IVB engines functioned properly and S-IVB restart was demonstrated satisfactorily. The O2H2 burner operated satisfactorily to repressurize the S-IVB fuel tank.

A free return translunar injection was obtained after S-IVB second burn cutoff. After spacecraft separation, the propellant vent, dump, and Auxiliary Propulsion System (APS) ullage burn successfully placed the S-IVB/IU/LTA-B in the slingshot trajectory to achieve solar orbit.

SPACECRAFT STRUCTURAL AND MECHANICAL SYSTEMS

All structural and mechanical systems performed satisfactorily, with the exception of spacecraft window fogging. The hatch (center, or #3) window was completely fogged over by about 6 hours. The two side windows (1 and 5) were similarly affected, but to a lesser degree. The rendezvous windows (2 and 4) remained clear throughout the flight. This fogging was consistent with what was expected as a result of the Apollo 7 analysis of window fogging, which was caused by a deposit of silicon oil on the inner surface of the outer heat-shield pane. The fogging results from the outgassing of the RTV compound which seals insulation around the window area. A curing process has been used on the compound in windows 1,3, and 5 of Apollo 9 and subsequent spacecraft.

THERMAL CONTROL

Temperature measurements indicate that both passive and active thermal control elements performed satisfactorily. Passive thermal control during the translunar and transearth coast periods stabilized spacecraft propellant temperatures within the expected nominal range. Tank temperatures were maintained within limits by varying spacecraft orientation. All temperatures were within predicted limits during lunar orbit operations.

FUEL CELLS AND BATTERIES

Fuel cell performance was excellent and no anomalies were observed throughout the mission. All parameters were in good agreement with preflight predictions.

The entry and post-landing batteries and pyrotechnic batteries performed all required functions satisfactorily throughout the mission. The entry batteries could be charged to full capacity as required.

CRYOGENICS

The cryogenic storage system performed satisfactorily throughout the mission, and usage was slightly less than predicted. All heaters were operated automatically and all fans were cycled manually to preclude the ac bus voltage problem caused by the arcing motor switch noted on Apollo 7. Quantity balancing between the respective cryogenic tanks was satisfactory.

COMMUNICATIONS

The overall performance of the spacecroft-to-network communication system was nominal. The received downlink carrier power, telemetry, and voice performance corresponded to preflight prediction.

Communications system management, including antenna switching, during the mission was very good. Communications during passive thermal control were maintained by sequentially switching between the four omni (omnidirectional) antennas, switching between diametrically-opposite omnis, or switching between the high-gain antenna and one or more omni antennas. All four omni antennas and the high-gain antenna were selected periodically, with performance equal to or greater than preflight predictions. The voice quality, both normal and backup, received throughout the mission was excellent. All modes of the high-gain antenna were used successfully.

The data quality of both high and low-bit-rate telemetry was good. High-bit-rate telemetry was received through the 85-foot antennas at slant ranges of up to 160,000 nautical miles while the spacecraft was transmitting on omni antennas. The MSFN sites reported receipt of good-quality telemetry data during data storage equipment dumps.

Communications were satisfactory during entry until blackout. Air-to-ground voice contact was re-established at approximately 146:52 GET through the Apollo Range Instrumentation Aircraft. The USS YORKTOWN established voice contact during parachute descent. Although post-landing voice communications were momentarily interrupted when the spacecraft was in a Stable II flotation attitude, the recovery operation was satisfactorily supported.

A total of six television transmissions were made during the flight. For the first telecast, the telephoto lens (100 mm focal length) was used to view the earth. Because of camera motion and the higher than expected light intensity of the earth, the pictures were of poor quality. A procedure for use of the filters from the Hasselblad camera was developed and subsequent telecasts of the earth using the telephoto lens with a red filter were satisfactory. Excellent views of the lunar surface were taken in lunar orbit using the extra-wide-angle lens (9mm focal length) and suitable filters.

INSTRUMENTATION

The data storage equipment and instrumentation system performance was satisfactory throughout the mission, and only three measurements failed.

The fuel cell 2 radiator-outlet temperature indicated a temperature 6 to 12 degrees higher than expected. Proper system performance of the fuel cell and radiator was verified by other system measurements, thus indicating a failed sensor.

The radiator-outlet temperature measurement in the environmental control system failed at approximately 120 hours elapsed time and went to full-scale reading. Systems analysis verified proper radiator operation.

The measurement of potable water quantity in the ECS failed at approximately 144 hours elapsed time. Normal tank pressurization and water production data indicated the measurement to be questionable.

GUIDANCE AND CONTROL

Performance of the guidance and control system was excellent throughout the mission. All monitoring functions and navigation comparisons required during ascent, earth orbit, and translunr injection were normal. Platform alignments were performed during all coast phases with good results. Onboard midcourse navigation techniques were thoroughly exercised. Star-horizon measurements were made during translunar and transearth coast, and preliminary comparisons indicate close agreement with ground tracking. Onboard orbital navigation was performed in lunar orbit with nominal results.

Spacecraft attitude control was satisfactory using both the digital autopilot and the stabilization and control system. Service propulsion maneuvers were performed using the digital autopilot, with nominal results. Entry guidance and navigation was excellent.

Two guidance and control system problems occurred during the mission. The first involved abnormal shifts in the computer readout of the optics trunnion angle. Several times during periods of no optics activity, the read-out shifted from 0 to 45 degrees. In each case, the correct reading was restored with a normal optics zeroing procedure and no optics utilization capability was lost.

REACTION CONTROL SYSTEMS

All command and service module reaction control system parameters were normal throughout the mission. The unplanned second CSM/S-IVB separation maneuver (see first item under FLIGHT CREW) resulted in lower than planned SM RCS propellant reserve, but still within acceptable limits. Due primarily to the need for only three of the seven scheduled midcourse corrections, the SM RCS propellant reserve converged on the planned curve for the remainder of the mission.

SERVICE PROPULSION SYSTEM

Four maneuvers were accomplished using the Service Propulsion System (SPS) and system operation was satisfactory in all cases. All maneuvers had "no-ullage" starts. The longest SPS burn was the 246.9-second lunar orbit insertion maneuver and the shortest was the 2.4-second first midcourse correction.

Early in the first SPS maneuver, a momentary drop in chamber pressure was experienced which was attributed to the presence of a small helium bubble in the oxidizer feed line. This bubble resulted from an inadequate bleed of the oxidizer system during preflight servicing. The chamber pressure was satisfactory throughout the remainder of the burn and for the three subsequent SPS maneuvers.

ENVIRONMENTAL CONTROL SYSTEM

Performance of the environmental control system was satisfactory. The radiators satisfactorily rejected the spacecraft heat loads during the translunar and transearth coasts, maintaining water/glycol temperatures below the evaporator turn-on level. The evaporators were therefore turned off during this time. The primary evaporator was used in the automatic mode during lunar orbit. Evaporator dryout occurred several times; however, the dryout did not impose any restraints on the mission. Evaporator dryout occurred on Apollo 7 at low loads and was expected to occur on Apollo 8 under similar load conditions. The evaporator was reserviced at the end of the first lunar orbit and operated satisfactorily until evaporator dryout recurred during the fourth lunar orbit. The evaporator was again reserviced and operated satisfactorily for the remainder of lunar orbital flight. Primary evaporator dryout occurred again during entry; however, the crew activated the secondary coolant loop, which operated properly throughout entry and maintained normal cabin temperatures near 61°F and the suit heat-exchanger outlet gas temperatures near 44°F.

The cabin fans, which were not needed during the mission, were noisy when activated at approximately 127 hours elapsed time to circulate the cabin atmosphere for a cabin temperature reading.

CREW PROVISIONS

All crew equipment operated satisfactorily during the mission. Excessive noise on the Lunar Module Pilot's electrocardiogram was corrected when he swapped the leads on his harness. The astronauts' boots were reported to be frayed but usable.

FLIGHT CREW

Despite the long duty hours, crew performance was sharp throughout the mission, and many valuable observations of the lunar surface and its environment were made.

The Apollo 8 mission was accomplished essentially in accordance with the nominal flight plan, with the following minor exceptions.

The S-IVB separation rate from the spacecraft was judged to be less than predicted, and the crew spent a longer time in keeping the S-IVB in sight and eventually used an additional Reaction Control System maneuver to increase separation distance.

Because of the heavy workload in lunar orbit, the orbital activities after the eighth revolution were sharply reduced to allow additional crew rest. Normal activities were resumed in preparation for the transearth injection, after which the flight plan was again modified to allow for additional crew rest. At about 100 hours the mission returned to the nominal flight plan with only minor rescheduling of rest and meal periods.

Entry and landing were performed in darkness with no apparent problems. Due to a small amount of water entering the command module at splashdown, the crew's attention was diverted temporarily. The resulting delay in releasing the main parachutes caused the command module to assume a Stable II (apex down) flotation attitude for about 4.5 minutes before being uprighted by the crew. A decision had previously been made to delay the deployment of swimmers until daylight; therefore, crew transfer to the prime recovery ship by helicopter occurred about 80 minutes after landing.

MISSION SUPPORT PERFORMANCE

FLIGHT CONTROL

Flight control support was excellent during the Apollo 8 mission.

NETWORK

Network performance was excellent for this mission. All communications, tracking, command, telemetry, and the real-time computation functions supported the mission satisfactorily with no significant loss of data at any time.

RECOVERY

Recovery of the Apollo 8 spacecraft and crew was successfully completed in the Pacific Ocean by the prime recovery ship, the USS YORKTOWN. The major recovery events on 27 December 1968 are listed in the following table:

EST, HR:MIN: (27 December)	EVENT
10:41	First visual sighting of spacecraft by Rescue I
10:43	Radar contact by USS YORKTOWN
10:49	First sighting of CM flashing light by YORKTOWN
10:52	Landing
11:48	Flotatian collar installed and inflated
12:20	Astronauts onboard recovery ship
13:18	CM onboard recovery ship

Both S-band and VHF contacts were established with the recovery forces. Visual contact with the flashing light and voice contact with the flight crew ceased at landing, indicating that the Command Module went into a Stable II position before uprighting. The uprighting bags were inflated, with one bag reported to be only partially inflated. Although a recovery helicopter was directly over the spacecraft as early as 11:08 EST, it was decided to wait until daylight before deploying swimmers, as previously planned.

The pertinent location data for the recovery operation are listed below:

Predicted target coordinates	08°08'N, 165°02'W
Ship position at landing*	08°09.3'N, 165°02, 1'N
Estimated range to spacecraft at landing	5200 yards
Splashdown coordinates	08°07.5'N, 165°01.2'W

* As determined aboard the recovery ship

LAUNCH COMPLEX

Quick-look assessment indicates that the launch complex and supporting GSE performed satisfactorily with only minor anomalies occurring.

Damage to GSE and pad facilities was minimal and considered less than for previous launches. Refurbishment will not be extensive and Pad A will be ready to support the Apollo 9 mission.